ENERGY CONSERVATION ISSUES IN THE U.K.

A Study and Commentary on the Administration, Power Structures and Politics of Energy Conservation in the United Kingdom.

by

Robert J. Jones.

Marine Marketing,
18, Southborough Road,
Chelmsford, CM2 0AQ,
United Kingdom.

First Published 1995.

ISBN 0 9519172 5 0
"Energy Conservation Issues in the UK."

British Library Cataloguing-in-Publication Data.
A catalogue record of this book is available from the British Library.

ENERGY CONSERVATION ISSUES IN THE U.K.

A Study and Commentary on the Administration, Power Structures and Politics of Energy Conservation in the United Kingdom.

CONTENTS

FOREWORD

This publication crystallises the lessons learned from the greater part of a working lifetime involved in the promotion of energy conservation.

Robert Jones has a perspective to bring to the topic. Working until 1993 as British Gas' co-ordinator for energy conservation, has given him a unique first-hand insight into so many aspects and activities.

The text covers in impressive detail each of the main areas of contention, from a coolly analytical standpoint. This is combined (most refreshingly) with insights into the personalities and institutions that have had the greatest impact upon the crusade for energy conservation, insights that can only have been gained by someone intimately involved with so many of the key initiatives.

Robert Jones' encyclopaedic knowledge and impressive memory for details succeeds in placing into context so many of the apparently random incidents which have bedevilled progress over the years.

If much of the book seems to be a catalogue of wasted opportunities and 're-inventing the wheel', then so be it. That is undoubtedly the way it was. And still is.

He rightly emphasises the confusion caused by the perpetual desire of Government to obtain energy conservation 'on the cheap', seeking continuously to rely for funding upon fuel suppliers whose vested interest it has long been to encourage, not discourage, consumption. He provides us with a litany of case histories of excellent initiatives permitted to peter out for lack of resources and/or political clout.

The book devotes considerable space to the chequered history of energy conservation education and training, a Cinderella area too often ignored. He pulls no punches, names names, gives praise where it is due, brickbats where they are due.

His conclusion is perhaps the greatest surprise of all. After such a catalogue of failure, his continuing optimism on progress - coupled with his prediction of the next big issue which will drive energy conservation -

makes this book well worth more than just a random dip. It deserves reading cover to cover, even if you are not an energy conservation enthusiast. As a catalogue of how public administration works - or in many cases fails to do - it is a unique document cataloguing our times.

Andrew Warren,
Director,
The Association for the Conservation of Energy.

1 INTRODUCTORY NOTE

In 1978, when "What's What in Energy Conservation" appeared, as a general over-view of the politics and organisation of energy conservation, there was a widespread acceptance that energy conservation was needed and that its promotion was a legitimate activity of government. Industry, commerce, academia and a wide range of voluntary bodies rallied around the "Save It" call and the U.K. began to enjoy something of a world reputation for leading the field.

At that time, the driving force behind energy conservation was simple and easily understood: Energy sources, as finite resources, were running out. If we did not save it, we would run out of it and civilisation as we knew it would end ! Around 1980 a new slant on the matter appeared: Even if we could get the energy we required, we would probably find we could not afford it. At the same time, a change of political attitude had us talking about "*cost effective* energy conservation", in response to a new political correctness that also sought to excuse industry from really doing anything serious about what the "Save It" slogan demanded. All the steam went out of the U.K. energy conservation programme. Even the word "conservation" became politically incorrect. "Energy Efficiency" replaced it and those who made profits from fuel and energy supply breathed sighs of relief. Such activity as remained was largely cosmetic and apologetic. Elsewhere in Europe governments continued their programmes and even Tory Energy Minister, Peter Walker, declared that the U.K. had slipped behind its European partners in energy efficiency. Then "the Environment" appeared.

During the "Save It" years some activists had tried to draw attention to the environment-damaging effects of profligate energy use but it took Global Warming and holes in the ozone-layer to bring this aspect to the fore. Although acceptance of the need for environmental protection on a broad scale was the initial driving-force, it was not long before the "Environmentalists" realised that energy efficiency and fuel conservation were the activities that would deliver environment-saving effects. Energy conservation re-emerged in environmental trappings. Thanks to fears about irreversible and disastrous damage to the environment, energy conservationists found themselves invited back onto the stage to help the U.K. politicians, business managers and commentators. They had re-discovered energy efficiency. Fuel conservation was back on the political agenda.

In summary, in the Seventies, energy was running out. In the Eighties, it would be too expensive, even if you could get it. "Use Wisely, not Do Not Use", became an acceptable refrain for the energy suppliers to sing. In the Nineties, energy use was seen to be destroying the environment. And that is where we are now. However, the lack of official commitment from '79 onwards has done sufficient damage to the energy efficiency/ conservation movement that the U.K. has lost its lead-position and has probably lost vast know-how and equipment export opportunities that may be irrecoverable, even in business horizons of decades.

This publication charts the history of the lost opportunities of the U.K.'s energy efficiency decline in an attempt to explain how the vestiges of the programme work - or fail to work - and how the decision-making process may be accessed or influenced. The picture that emerges is not encouraging.

For the future, a whole new generation of enthusiasts will be needed because it is now over a quarter of a century since the "Save It" activists learned their trade and earned their spurs. If there is hope for that ageing band and for their emerging successors, that hope itself is based on a distressing premise. There is a growing awareness that there is a new imperative to add to the old list of scarcity, cost, strategic economic security and damage to the environment. That sinister imperative is the realisation that energy-use is killing us. Researchers are beginning to understand the mechanisms of the killing process, one of which is that emissions from fossil-fuel burning attack the immune systems of all forms of life - even the life of humans. That, as the popular press might say, is a whole new can of worms. It is certainly a grisly thought but it could be the next reason why energy conservationists are asked to dust-down their books. As was said so often about the problem, even back in the Seventies: "We know what to do. The problem is getting people to do it". If this publication shows where and how decisions are made, or where pressure may be applied to influence policy in energy conservation, then something may be done - this time round.

Apart from responding to encouragement to up-date the 1978 study, I was setting about this book whilst being involved in staff selection procedures and research grant-making decisions for energy conservation organisations and university faculties. These processes indicated that education and general awareness in the energy conservation field are deficient. The technicalities are fairly well taught. The micro- and macro-economics are generally well understood and extensively researched. What is lacking is a knowledge of what has happened, how and why it happened and how the machinery of energy conservation decision-making works. Above all, as in any human activity, there are profound political forces to consider. In energy conservation they are formidable and need to be recognised. I hope this study also adds a little understanding to recognition.

In the 1978 book, something of an artificial boundary was drawn around 'energy conservation', excluding many other energy issues. This approach is repeated, even though the artificiality is increased by the development of environmental issues since 1978. Energy conservation is certainly only part of the wider energy policy and environmental scene but any writing must set itself a boundary - if only to save trees!

Some years ago I had the good fortune to be delayed at a Scottish air-field in the company of Sir Alex Douglas-Home. He remarked on the volume of Winston Churchill's, "Second World War" that I was reading and told me that he had said to the Great Man that it was a good read but very biased. "Yes", agreed Churchill, "but it was great fun writing it". I think this little venture has led me to appreciate what he meant!

Robert J. Jones,
Chelmsford 1995.

ENERGY CONSERVATION AND GOVERNMENT

"The energy conservation action scene is both complex and complicated, with apparent duplication, over-lapping and sometimes confusing lines of responsibility or influence between government, quasi-government, professional and voluntary bodies. Even completely non-statutory professional and voluntary bodies manage to aquire special standing by becoming associated with statutory organisations and government departments."

"What's What in Energy Conservation", RIBA Publications 1978.

2.1 WHO 'OWNS' ENERGY CONSERVATION?

2.1.1 The degree of complexity of the energy conservation scene has certainly not diminished since the late Seventies, though it is now a complexity born of neglect rather than of the enthusiasm for action of earlier times. In 1978 one was able to write such statements as, "As far as Government itself is concerned, there is the Energy Conservation division of the Department of Energy at the top of the pile...." Now it is not clearly evident that the "pile", if it has a discrete identity, has only one "top", nor what the pile is supposed to be for, nor who really owns it.

2.1.2 The casual observer coming upon the energy conservation scene in the U.K. is first confronted by three official sounding names : The Energy Efficiency Office (EEO), The Energy Saving Trust (EST) and The National Energy Foundation (NEF). Only one of these, the EEO, is in fact "official". Although the Trust (EST) was a creation of Government, Ministers and Officials are at pains to explain that it is not an organ of Government, that it has a life of its own, is independent and so forth. This position is maintained even though Government is one of the "Trustees" and even though the Trust administers a government-funded initiative, the local Energy Advice Centres (LEACs) (Chapter 6 and Chapter 9). The Trust is also presented as one of the most important elements of the Government's Climate Change commitments. On the other hand, the National Energy Foundation is genuinely independent of Government, in origin and in direction. However, it does manage for the EST the LEAC scheme, that is part-funded by Government, and which the EST administers for EEO; itself part of a Government Department! ("The energy conservation scene is both complex and complicated......")

2.1.3 Until April 1992 there were three distinct "piles" of Government activity in energy conservation, contained within three Departments of State: The Department of Energy (DEn), The Department of the Environment (D.O.E.) and The Department of Trade and Industry (D.T.I.). (Strictly speaking, the Department of Transport could be viewed as a fourth player but its contribution to energy conservation has been minimal, even though transport accounts for around 30% of energy use and is the one energy-use sector that appears to be totally out of control, with energy consumption growing at a ludicrous rate. Raising this alarming situation with D.En, DOE or DTI would elicit merely a curt response that it is a Dept. of Transport matter.) In 1992 the Dept. of Energy was dismantled and its Energy Efficiency Office transferred to the D.O.E. . The EEO did not move cleanly, as a unit, into its new home; some of its responsibilities were taken up by D.T.I.

2.1.4 Relations had always been difficult between D.O.E. and D.En. over energy conservation, so the Officials of E.E.O. were not over-joyed by the move; though the more career-minded were pleased to be moving into a larger promotional pond. What was most extraordinary about the move was that, instead of the E.E.O. being an integral part of Environmental Protection, by 1994 it found itself subsumed into the Housing Directorate. Incredibly, this was said to be done chiefly to even-up staff numbers. If ever such a signal was needed, this was a clear signal that energy conservation was not high on the agenda of either Whitehall or Westminster. Considering that housing represents only a modest percentage (34%) of the energy demand issue, putting the E.E.O. into the D.O.E.'s Housing Directorate could only be viewed as a live-run test for a new series of, "Yes, Minister".

2.1.5 This extraordinary move should not have been a surprise. In 1991 Government, in the persons of Michael Heseltine and Peter Lilley, set up a new advisory committee, the Advisory Committee for Business and the Environment (ACBE) as a joint D.T.I./D.O.E. initiative. The D.En. and its Energy Efficiency Office were not involved and only crept in, thanks to external lobbying by energy conservation pundits, through a sub-committee, the Global Warming Working Group. Once in, they made a good contribution but the situation only seemed to suggest that D.T.I. and D.O.E. held the E.E.O. in little more than contempt. The curtailment of energy use as a key to environmental protection appears to have escaped the political and administrative minds. However, the non-government and non-civil service Members of the Committee were quick to grasp the point and report forcibly on it.

2.2 THE ENERGY EFFICIENCY OFFICE NOW

2.2.1 In spite of the arcane arrangements of Whitehall, the newcomer to energy conservation in the U.K. would reasonably expect to start with the Energy Efficiency Office in searching for the activating levers in this particular aspect of the machinery of government; so it is useful not only to fix its organisational position, but also to establish its relative political importance.

2.2.2 As explained, with the demise of the D.En., the E.E.O. became a part of D.O.E. . The D.O.E. itself is organised as a number of Commands, each headed by a Deputy Secretary (Grade2), roughly one down from Humphrey-level of "Yes, Minister". These divisions cover such areas as Planning, Rural Affairs and Water, Local Government, Environmental Protection, Property Holding, Housing and Construction, Regional Organisations and so forth. The E.E.O. is presently part of the Environment Command and is headed by what the Service terms a "Grade 3" . (Grade 1 is the top!).

2.2.3 The head of the Energy Efficiency Office is currently styled, "Director General". In D.En. days this title was adopted to indicate the out-reaching nature of the E.E.O. but it does not seem to be widely bandied about these days. In 1992, when the Office was absorbed by D.O.E., was it portentous that "Vacher's Parliamentary Companion", with rare inaccuracy, relegated its name to "Energy Efficiency Unit", reminiscent of the original humble name, "Energy Conservation Unit", of its earliest days?

2.2.4 The E.E.O. nominally has four Directors but, at the time of writing, only three of these report directly to the Director General : The one heading Environmental Protection and Industry (EPI) has a sort of dotted-line relationship to the head of the Environmental Protection Directorate. This is not entirely surprising because his remit includes the Advisory Committee on Business and the Environment (ACBE), which is "owned" by that Directorate. As if to stress this semi-detatchment, it is noticeable that, of the four Directors, this one is known as "EPI" whilst the others are unoriginally, but tidily, known as "EEO1", "EEO2" and "EEO3". Each is headed by a Grade 5.

2.2.5 EEO 1 (Domestic Sector and Marketing) lists its responsibilities as:

- Domestic buildings publicity campaigns.
- Low income households.

- Home Energy Efficiency Scheme (HEES).
- Local Energy Advice Centres (LEAC) pilot scheme.
- Marketing policies.
- Regional energy efficiency network.

2.2.6 EEO 2 (Non-Domestic Sector and Policy) has a more ill-assorted remit. The individual elements tempt the observer to raise such questions as, 'What is Policy?' or, 'What has home energy labelling to do with the non-domestic sector?'

- Public Sector, Industry and Commerce.
- Policy and Finance.
- Home Energy Labelling.
- Education and Training.
- Energy Conservation Bill.
- Energy Saving Trust.
- Energy Management Assistance Scheme (EMAS) - recently terminated.
- International Issues.

2.2.7 EEO 3 (Technology) presents an equally disparate picture, suggesting that it has the most mis-leading title of all; leading one to wonder why, "Appliance Labelling" is 'Technology' whilst, "Home Energy Labelling" is "Non-Domestic Sector and Policy". It lists its responsibilities as:

- Best Practice Programme.
- Appliance Labelling Initiatives.
- Management of the Energy Technology Support Unit (ETSU) and the Building Research Energy Conservation Support Unit (BRECSU).
- Promotion of Combined-Heat-and-Power (CHP).
- Environment Technology Innovation Scheme.

At the time of writing (1994) the EEO 3 was due to go, on the occasion of the retirement of its Director. The activities will be redistributed. (Thus saving one job!).

2.2.8 The stylistically odd-man-out, EPI (Environmental Protection and Industry) has a list that explains why it leans organisationally towards the Environmental Protection Directorate:

- Advisory Committee on Business and the Environment (ACBE).
- Promotion of Positive Environmental Management in Industry.
- EC Eco-Management and Audit Regulation, BS 7750.
- Green Consumerism and Ecolabelling.
- Making a Corporate Commitment Campaign (MACC).

2.2.9 These divisions of responsibility change from time to time but the above analysis does at least demonstrate that care is needed when seeking out which part of the EEO is interested in a particular aspect of energy conservation. It is particularly important to note that EEO 3 (Technology) manages the Building Research Energy Conservation Unit (BRECSU) and the Energy Technology Support Unit (ETSU). These are, respectively, part of the Building Research Establishment (BRE) and of the Harwell Research Establishment. BRECSU chiefly provides technology backing on buildings and domestic energy use, whilst ETSU is chiefly concerned with industrial process energy use. Each of these units is bigger, in staffing terms, than EEO itself and, besides technology competence, produces excellent energy conservation advisory material, seminars and conferences. However, it is important to remember that both BRECSU and ETSU service the EEO and that it is within the Energy Efficiency Office that the decision-making power lies concerning programmes and priorities.

2.3 OTHER OFFICIAL 'HOMES' OF ENERGY CONSERVATION

2.3.1 Where buildings are concerned, probably the most powerful instruments for achieving energy efficiency and fuel conservation are the Building Regulations. These, of course, are concerned with wider issues than energy use; so it is to be expected that responsibility for them lies outside the EEO, in the PBD (Property and Buildings Directorate) of Housing and Construction. Within PBD there is a section called "BR(Building Regulations)" and it is there that energy conservation regulation resides. (PBD also owns a section known as PASD (Property Advisory Services), which lists, "Conservation Unit", under its responsibilities. This, however, is not energy conservation, rather heritage and preservation of listed buildings.)

2.3.2 One other directorate with an interest in energy conservation is HPPF (Housing Policy and Private Finance) which, in its HRP (Private Sector Renewal Policy), lists, "Energy Efficiency in Housing" and "Housing and Environment".

2.4 OFFICIAL SOURCES OF ADVICE

2.4.1 To carry out their duties, Ministers of State seek advice beyond that offered by their civil servants. Up to 1981 there had been a well-established advisory body, the Advisory Council for Energy Conservation (ACEC). In 1981 it was disbanded and the Prime Minister herself appointed five "Marketeers" in 1983 to take over the energy conservation advisory task. ACEC had achieved much and published many substantial reports on the subject; both as direct advice to policy-makers and as operational advice, but it is difficult to determine what the new quintet achieved collectively; though one of them had made significant contributions to the EEO's marketing programmes and two had tried to get movement on the energy labelling of homes when they were first appointed.

2.4.2 In 1990 an advisory committee on the environment was established which, by definition, would include advice on energy conservation. This was a joint creation by the Secretary of State for the Environment (Michael Heseltine) and the Secretary of State for Trade and Industry (Peter Lilley) and was titled, the Advisory Committee on Business and the Environment (ACBE). Whether an advisory committee is a lesser creation than an advisory council is not generally clear. It certainly sounds less impressive but, equally, it sounds more impressive than a panel; such as the Energy Advisory Panel (EAP), that appeared in 1993 (see below).

2.4.3 Significantly, but not surprisingly, none of the "Five Marketeers" appeared on ACBE, whose membership was drawn widely from leaders of industry and commerce, plus those experienced in environmental protection, waste recovery and energy efficiency. Its Chairman was John Collins, of Shell U.K., who received a well deserved knighthood for his efforts plus some bemused reactions from colleagues in the oil industry, who suspected that he had "gone native" with some of the recommendations of his committee. In fact, his stance indicated that there are some in U.K. industries who are sufficiently well informed, experienced and public-spirited to be able to rise above short-termism and sectional interest. ACBE produced a series of reports for Government which were considered and won detailed reactions from ministers, who took aboard a number of the recommendations. There were also two publications on energy efficiency for industry and commerce, covering buildings and transport, the former having become something of a text-book for energy managers. ("Practical Energy Saving Guide for Smaller Businesses" - E.E.O.). These and the earlier reports of ACEC make excellent reading for those seeking an introduction to energy efficiency and environmental protection issues as viewed by industry and commerce. Most of the ACEC reports were published in the Government's "Energy Papers" series by HMSO.

2.4.4 An innovative feature of ACBE was that its members voted for disbandment after their initial two-year appointment period and advised the Secretaries of State to form a new group, if they wished, but not to prolong the life of the original body. In the Spring of 1993 it made its final report and in the Autumn of that year a new group was formed under the Chairmanship of Derek Wanless, Chief Executive of National Westminster Bank, plc., and a member of the former group. It is an interesting reflection on the challenges that environmental issues appear to imply for industry and commerce that an eminent banker takes such a prominent position. However, the problem for business from environmental issues are more likely to arise from the implications of contaminated land liability rather than from energy conservation itself.

2.4.5 In 1993 a further advisory body was set up by the Secretary of State for Trade and Industry, which can be expected to have a bearing on energy efficiency. This is the Energy Advisory Panel (EAP); though, as its title implies, its remit is on energy policy in general rather than on energy conservation in particular. It was set up following the Coal Review White Paper and is chiefly to advise the government on its annual, "Energy Report". At the same time, the creation of the Panel was taken as an opportunity to disband the Advisory Council on Research and Development (ACORD), which had provided an overview of the research programmes of the publicly-owned energy industries for over 45 years - programmes that had included substantial work on energy efficiency. Only two out of the Panel's sixteen members, at the time of its inception, had any experience or expertise in energy efficiency; so it will be interesting to see whether it operates chiefly for the supply-side or the demand-side of energy policy

2.4.6 That energy conservation advice to ministers is now coming from advisory groups set up with the wider remit of 'environment' merely reflects the change of public and official emphasis. As mentioned, there was an element of surprise that the main thrust of ACBE's advice concerned energy conservation, when it had been set-up under an "environmental" umbrella and had had what appeared to be a predominantly waste recovery and pollution clean-up agenda. The Government's Global Warming Consultation Conferences in 1992 made a similar point of departure but moved chiefly into an energy conservation mode. It is inescapable that most environmental pollution and global warming problems come back to energy-use for solution.

2.4.7 In 1995 two further advisory groups appeared that will be feeding into official policy; again, each has an environmental title but both will certainly be addressing energy efficiency and fuel conservation issues, if they are to offer anything of real value. The U.K. Round Table on Sustainable Development has been set up by Government in the hope of adding impetus to activity on environmentally sustainable development, mainly looking to the long-term. Its membership, of about 30, is drawn from business, environmental and related interests. It has a five-point agenda:

- help identify the agenda and priorities for sustainable development
- develop new areas of consensus on difficult issues of sustainable development
- provide advice and recommendations on actions to achieve sustainable development
- evaluate progress towards objectives
- inform and involve others, building wider support for the issues

The problem with this sort of agenda is that it implies an operational role, suggesting an attempt to displace civil servant resources with unpaid committee people. Advisory bodies should keep to advice.

A further new body, the Renewable Energy Advisory Committee, replaces the Renewable Energy Advisory Group and advises the Dept. of Trade and Industry on its renewable energy programme, "particularly with regard to the aims, content, environmental appropriateness and economic applicability of the programme and policy on new and renewable energy". It comprises industrialists, scientists, economists and energy scene activists. It is on the supply-side of energy efficiency and may only have passing interest for energy conservation.

2.5 U.K. GOVERNMENT PROGRAMMES

2.5.1 It is a sad fact of life that energy conservation is no longer high on the U.K. Government's agenda. Its diminished place in the machinery of government reflects this. In the 1994 re-shuffle there did re-appear a junior ministerial post that could be termed, "Minister of Energy Conservation" but his main tool for action, the Energy Efficiency Office, continues to shrink in staff and budgetary provision, unless one is prepared to accept the increasing HEES budget as a legitimate 'energy conservation' fund. Once the provision for the Home Energy Efficiency

Scheme (HEES) and its Northern Ireland equivalent DEES (Domestic Energy Efficiency Scheme) is taken away (It is, after all, a social security rather than an energy-saving provision) there is little left. Energy Conservation observers should be wary of claims about government funding that include HEES and DEES, when official sources are trying to imply a real commitment to energy conservation.

2.5.2 In defence of government, the observer could pick over the various detached and semi-detached organisations that would, in pre-privatisation and pre-agency-mania days, have been organs of government and have "counted" as official financial commitment to energy conservation. The Energy Saving Trust (Chapter7) is the most obvious of these and the 160-plus staff of the Energy Action Grants Agency (Chapter 10) would, in more reasonable communities, be direct government employees if one accepts, of course, that HEES has some conservation significance. On the down-side, it has to be remembered that government expenditure through these channels is subject to various layers of "top-slicing" and profit-taking, before anything is actually spent on saving energy. The Government's Local Energy Advice Centres (LEACs) project is a particularly multi-layered and salutary example: Government contracts with the Energy Saving Trust to direct the initiative. The Trust takes a "slice", believed to be between 10% and 15%, and then contracts with the National Energy Foundation to manage it and develop an I.T. capability. The Foundation takes its fee and then contracts-out the I.T. job and invites bids from various organisations to run the centres. These organisations, ranging from Fuel Poverty charities to a Regional Electricity Company, in turn take their fee or "profit" and, at the end of this lucrative line, a few callers receive advice on how to save energy in their homes. The cost-per-kilowatthour-saved ratio, when the initiative is finally audited, will be interesting and, no doubt, instructive. Figures around £75 per call are being circulated for the cost to the Public Purse of each advice-seeking telephone call attracted by the LEACs but this is merely hearsay until the LEACs first-year accounts are published.

2.5.3 Besides the privatised, contracted-out or agency services, the machinery of government for energy conservation has also to consider the various fuel supply regulators and officially-funded consumer organisations that have a say in what is or is not done and how it is or is not funded. These bodies have been of mixed use to the cause of energy conservation. Offer (The Office of Electricity Regulation) has begun to put demand-side control duties on the Regional Electricity Companies but, so far, involving trivial sums. Ofgas (The Office of Gas Regulation) started down the same route; in fact, being largely responsible for enabling the Energy Saving Trust to come into being. However, a change of Regulator brought that to a halt and the actions of the new Regulator have, almost single-handed, brought the Government's much-vaunted Climate Change programme to its knees. In general, the place of energy conservation in the process of regulating the former publicly-owned fuel utilities emerges through the "Standards of Performance" (SoP) regulations. "SoP" is an extraordinary piece of bureaucratic/political double-talk; really meaning, "Public social services that would not be produced by market-forces". They are not standards; merely definitions of activity obligations. The consumer organisations are becoming increasingly critical of official inaction where energy conservation and environmental protection are concerned; so it is likely, in the Thomas à Becket world of current U.K. administration, that they may soon cease to be part of the machinery as we know it. Although their interests are wider, these organisations are interested in energy conservation when the subject affects the purses or even health of the consumer. Electricity Industry Consumer Councils did not survive privatisation of the Industry but the Gas Consumers Council did.

2.5.4 To the outside, the machinery of government for U.K. energy conservation may seem delightfully simple: A designated department of government, an Energy Saving Trust and regulators for the energy suppliers - or some of them! In reality, the regulators are either out of control or inclined to agendas other than energy conservation and climate protection. The Energy Saving Trust is starved of funding and expensively constituted. The designated department of government is a shrinking, marginalised unit in a backwater of the Department of the Environment. The result has been a steady decline in activity, a steady erosion of morale among enthusiasts and a worsening of the U.K. energy efficiency ratio.

2.5.5 In spite of the general run-down, there are certainly some excellent government-funded schemes and publications to encourage energy efficiency; either continuing or being developed. These are detailed in other sections. However, they chiefly reflect the pool of expertise that is surviving the overall run-down of the machinery and themselves suffer from under-funding or funding uncertainty. The support of some of the energy-suppliers has dwindled in response to fading official commitment and as a result of privatisation. This has further deprived the "official programme" of funding and expert help. The U.K. no longer has a Department of Energy. It will be interesting to see how long it retains its Energy Efficiency Office as more and more of an EEO's duties are contracted out. This process itself adds to the complexity of the U.K. official machinery for energy conservation, in obscuring what is "official", "un-official", "quasi-official", QUANGO or totally independent.

2.5.6 Within its general neglect of conservation, the U.K. Government does still enable some bright ideas to surface and does give "official backing" to them. The problem is that "official backing" rarely means funding or, if it does, rarely gives adequate funding or else shackles funding to total impractical mechanisms to disguise intervention as private sector initiative. The concept of "Energy Action Cities" is a good example of a useful energy conservation idea as yet another lost opportunity.

2.6 ENERGY ACTION CITIES/ENERGY CONSERVATION AREAS.

2.6.1 "Carrots and Sticks", has for many years been a well-used phrase to describe government action but it was during the Eighties that the energy conservation sector began to identify and articulate a third action option, which was, "the Tambourine". Governments might offer incentives in the form of grants, tax-breaks, subsidies and so forth ("the carrots") or impose penalties ("the Sticks"). "The Tambourine" is the process of encouragement by, for example, high-profile support, lending the government's name to a project or wheeling out a minister to give status and create media attention - rattling the tambourine to make others feel enthusiastic and jump through the hoop.

2.6.2 The Energy Action Cities initiative was an example of an energy conservation "tambourine"; encouraging local authorities and other bodies to generate activity in assessing the energy-saving potential of a locality but with little or no government financial or resource support - apart from lending such terms as "Government backed" and producing ministers and high-rank civil servants at launching or reporting events.

Besides "Energy Action City", there were also terms such as "Energy Conservation Area" or "Energy Project Area" applied to the idea and it is also difficult to be precise about when the initiative really began, who first conceived of it or how it became taken-up and re-packaged by government officials. There were certainly some locally-generated initiatives but similar concepts were already being articulated nationally by academics, professional bodies and energy conservation lobbyists.

2.6.3 In October 1982 the National Energy Efficiency Forum (NEEF) convened a conference, sponsored by the Dept. of Energy, to launch the concept of what they termed, "An Energy Conservation Area". The NEEF (A fairly informal group of energy conservation enthusiasts) was looking for a local authority to set up a pilot scheme with a view to replication, if it were successful. The idea was to create co-ordinated action, involving individual householders as well as collectives and businesses - a sort of critical mass of activity that would then generate further momentum towards establishing a low-energy community. Approaches varied from intellectual frameworks, involving macro surveys of energy use, to headlong programmes of energy conservation measures in all sectors, plus many combinations of these two stand-points.

2.6.4 There were over 50 participants at the conference, including representatives of fuel suppliers, conservation lobbyists, academics and local authorities. In March 1983 a conference report was published ("Blueprint for Action", published by the Consumers Association) but no positive action arose. However, at this point the Energy Efficiency Office took over the idea from NEEF and encouraged development. A scheme was started in Manchester, but on the basis of an energy audit of the area, and other county or regional programmes were attempted.

2.6.5 The term "Energy Action City" was not widely coined until about 1985. Although there had been central government money, to back local government funding, the organisers also turned to the fuel suppliers for assistance and were particularly reliant on them for energy audit data, where a scheme was to be based on a regional energy audit. This may have been one reason for the general lack of impact of the idea: The suppliers were not over enthusiastic about organising massive reductions in local energy markets. Even so, without even the small amounts of money forthcoming from the fuel supply utilities - and occasional sums from oil companies - it is doubtful whether any of the early ventures would have occurred.

2.6.6 By 1985 Cardiff was declared an Energy Action City and the EEO had received supportive responses from a number of other local authorities - among these, Newcastle-upon-Tyne, Bolton, South Glamorgan, Glasgow and Dudley, some of whom were well advanced with their own planning.

After Cardiff, a change of Secretary of State seemed to bring a reversal of official attitude towards the idea but it was revived in 1988 with an announcement that Bristol, Dundee and Norwich had been designated, with Edinburgh, Hull and Sterling known to be likely candidates. The EEO aim at that time was to have 25 cities designated and running by 31st. March 1988.

2.6.7 Overall, results have been patchy. Some reports have been produced but nowhere has the original comprehensive concept of the 1982 NEEF conference been realised. However, where initiatives have been followed through, there have been some notable localised achievements. Some ventures have become well established as on-going activity; for example, Norwich Energy Action was set up in January 1988 and was still discernible as a local movement at the time of writing - thanks largely to on-going support by the local Regional Electricity Company. Even so, it is far from being a locally managed low-energy community. The pity is that such local successes indicate how much enthusiasm and goodwill exists that could be made so much more effective with just a working level of financial help and official support. It is surprising that such a Right Wing administration has not acted on the old military rule: "Re-inforce success". This is a weakness of the present U.K. machinery of government that is manifested in many other energy conservation missed opportunities, from the mega-concept Energy Saving Trust to the local Energy Managers Groups.

2.6.8 The Isle of Wight was an extreme example of death by official neglect plus the irony that fuel privatisation measures also legislated the venture into failure.

Attempts had been made to establish the Isle of Wight as a sort of isolated regional exemplar for an energy conservation area - literally isolated by its geography but still a highly-developed and thriving economy. It also offered the prospect of European money because the Commission was, at that time, looking for a suitable island community for study, where the geography would enable a tight audit of energy input and energy use. In the absence of hard-cash U.K. government support - and, unfortunately, some personality clashes between tiers and locations of officials - the initiative foundered.

The failure also high-lighted an interesting fall-out from the energy industries privatisation process: The study would have needed detailed energy input figures from the energy suppliers; the electricity and gas utilities, plus oil and, to a lesser extent, solid fuel companies. In principle all were agreeable but then British Gas realised that it would be releasing commercially sensitive information concerning the psuedo-competitive market that government legislation and regulation was attempting to create. The Company refused to co-operate, so the project would not even have had the gas element of the energy data needed for its basic energy-supply audit. Having delivered the entire gas supply industry into private hands, the government could not do much to alter this intransigence. The Isle of Wight Energy Conservation Area was not established. Considerable local authority and local business enthusiasm was wasted.

2.6.9 The Energy Action City/Energy Conservation Area concept should not be confused with Inner City Initiatives and such ventures as the Green House Programme. These, indeed, had energy conservation elements and were sometimes part of Energy Action City programmes but were totally separate initiatives, with economic regeneration or environmental commitment agendas. Hopefully, the Home Energy Conservation Act of 1995 will create de facto Energy Action Areas everywhere; a rose by another name?

2.7 ENERGY POLICY AND THE POLITICAL AGENDA.

2.7.1 The casual observer might expect to find an appreciation and understanding of U.K. energy conservation within an overall "energy policy". Indeed, back in the Seventies it was commonplace that such a policy was encapsulated in the term, "Coconuke" - an energy policy based on coal, conservation and nuclear power. Since then it has become something of a tiresome sick joke that the U.K. energy policy is that it has no energy policy! There is certainly now no Department of Energy and government has little control over the fuel-supply industries, now all in the private sector - with the exemption of British Nuclear Fuels. Any driving force for energy conservation - of which there is very little - comes from Earth Summit commitments, rather than industrial security or economic survival. Even there, government has placed its hopes on two conspicuous failures: V.A.T. on fuel and the Energy Saving Trust. There is certainly no effort, such as was attempted in the Seventies with the Energy Commission, to bring all energy suppliers and users together to produce a national energy policy. Some pressure groups try to induce government to introduce "least-cost-planning", under various guises, to curb the pile-it-high-sell-it-cheap instincts of the energy producers. The Regulators of the privatised energy industries seem unwilling or unable to work outside maximising profits and controlling prices and are dragged reluctantly into even token energy conservation measures. Where oil product suppliers are concerned, there are no Regulators at all. The fierce competition that was waged between gas and electricity, that was largely destroyed by privatisation, had been the driving force behind their once extensive energy efficiency programmes. These have now mostly disappeared, along with funding and sponsorship for the initiatives of government and others.

2.7.2 In the absence of a national energy policy, it is not surprising that there is little evidence of a national energy conservation policy. It is equally not surprising that there is only a small and shrinking Energy Efficiency Office, that it does not have a direct interest in transport and that it still seems to sit uneasily within the Environment Protection Directorate and that the U.K's energy-to-GNP ratio is worsening.

The U.K. is in the enviable position of having abundant and varied energy supplies - for the time being. It has the luxury of fuel choice. It has energy management technology excellence and the capacity to reduce its energy use by at least 20% - some would say, 25% - without much effort or disturbance. The effect that that would have on the now privatised and increasingly internationalised energy supply industries' profits would be interesting! Perhaps that explains why energy conservation is so low on the political agenda. As in all human activity, in trying to understand how a system works, how the U.K. energy conservation system works, one needs to keep in mind the well-worn dictum, "Cui bono?" - "Who's making a killing?"

3 LEGISLATION AND REGULATION

3.1 In spite of plenty of advice to the contrary, government in the U.K. since 1979 has shown a marked reluctance to legislate to promote a low-energy society. The two welcome exceptions have been the 1994 improvements to the energy conservation aspects of Building Regulations and the quasi-legislative Standards of Performance in fuel supply regulation. In spite of official opposition to legislation, by 1995 even the Energy Saving Trust - of which Government is a Founder Member - was calling for legislation; this time to force the Gas Suppliers to fund energy efficiency projects.

3.2 This reluctance on the part of Government was manifested most forcibly in 1994, when a Private Member's Bill sought to introduce a new Energy Conservation Act. Its provisions were modest; merely to require local authorities to identify how to cut fuel consumption in domestic premises by 30%. It had wide cross-party support but was shabbily "talked out" by Government itself, using a procedural device of tabling 237 amendments at the very last stage of the Bill's passage through Parliament. It was re-introduced, again as a Private Member's Bill, in 1995 and was finally passed into law in the Spring. However, it is not the first of its kind. There had been a Homes Insulation Act in 1978 and an Energy Conservation Act in 1981.

3.3 THE 1981 ENERGY CONSERVATION ACT

3.3.1 In May 1981 an Energy Conservation Bill completed its passage through Parliament and became the 1981 Energy Conservation Act. Most of its provisions were concerned with giving powers to Government to set mandatory standards for energy efficiency and safety of all appliances designed for space-heating and hot water production, plus gas appliances used for cooking, refrigeration, lighting and washing. It enabled the Government to require type approval for appliances sold in large numbers and for on-site testing of very large boilers. Very little has come of this legislation and Government has not since used the powers.

3.3.2 The Act did, however, have a section, Part III, that was said to be merely a device to regularise the funding of certain schemes; such as, the Energy Quick Advice Service (EQAS), the Energy Survey Scheme, the

Extended Survey Scheme and so forth. (All of these schemes were, in fact, quickly wound-up as the new administration got into its stride in dismantling the energy conservation structures that had inherited in 1979).

The explanation at the time for this very interventionist piece of legislation was that hitherto the EQAS and similar schemes had had their funding legitimised rather obscurely and tenuously by a clause of the Fuel and Power Act of 1940. The new provisions of the 1981 Energy Conservation Act were to regularise the existing schemes; there being no intention at all to use them to start new services. However, some officials claimed that the Act could be used to give a Secretary of State very wide powers to use public money to promote energy conservation. The Act remains in force and the relevant provision states:

> "15.- (1) The Secretary of State may with the approval of the Treasury make grants in accordance with this section for purposes of any scheme for the provision of advice with a view to promoting the conservation of energy.
>
> (2) Grants under this section
>
> (a) may defray in whole or in part the cost of any of any particular premises, or of premises within any particular area, for the purpose of determining appropriate measures to be taken in the case of those premises for reducing the use of energy;
>
> (b) may be made with reference to advice relating to the use of energy for any purpose whatsoever (domestic, industrial, commercial or otherwise;
>
> (c) may be paid to the persons giving advice or to persons receiving it, or partly to the one and partly to the other; and
>
> (d) may be made subject to such conditions as the Secretary of state may determine."

3.3.3 The Act has enabled the various grant-making schemes to be introduced and could be used to fund practically everything that the Energy Saving Trust and others want to do to save energy. All that is required is political will.

3.3.4 The Homes Insulation Act of 1978, which enabled the Homes Insulation Scheme to be introduced, was an example of what could be done, if Government wished to make a serious in-road into the U.K.'s energy saving potential. Soon after it came into power, in 1979, the new administration set about cutting the £25 million being spent by the Act but re-instated it, at £27 million, only ten months later. In later years it was dropped completely and gradually replaced by a more targeted approach that exists now as the Home Energy Efficiency Scheme (plus the Domestic Energy Efficiency Scheme for Northern Ireland) - (See Section 8 - "Fuel Poverty").

3.4 THE 1995 HOME ENERGY CONSERVATION ACT

3.4.1 After the shameful Government performance over the demolition of the 1994 Energy Conservation Bill, it was a welcome relief that the next attempt, also a Private Member's Bill, the Home Energy Conservation Bill, successfully completed its passage through Parliament to become the 1995 Home Energy Conservation Act.

Like the former attempt, this Bill had all-party support and widely drawn public support. This was well demonstrated when the Second Reading stage was supported by a national rally and mass lobby of Parliament. Besides all-party MP participation, the rally had such diverse and nationally respected sponsorship as Help the Aged, the National Right to Fuel Campaign, Friends of the Earth, Age Concern, Neighbourhood Energy Action, Greenpeace, the Association for the Conservation of Energy and the two major trade unions, UNISON and NUS.

3.4.2 It is of interest that the Bill's sponsor, Diana Maddock, Liberal Democrat MP for Christchurch, had originally wished to call it, "The Warmer Homes and Energy Conservation Bill", to demonstrate its relevance to the Fuel Poverty aspects. However, Parliamentary draughtsmen objected on the grounds of subjectivity and imprecision.

A significant amendment was the addition of "advice" to the list of what constitutes an "energy conservation measure" within the Act but the sponsors of the Bill were disappointed that a requirement for prioritising action was dropped and that specific targets were omitted. However, in the course of the debate the Government spokesmen affired an intention to set individual targets for local authorities according to past achievement and revealed need.

3.4.3 The essence of the Act is that local authorities will be required to prepare Energy Conservation Reports. These will be submitted to the Secretary of State for the Environment and will be published. The

Secretary of State will approve Reports and require progress reports on progress in implementation. This will include setting targets and giving guidance. In turn, the Secretary of State is required to report to Parliament on the overall progress.

An Energy Conservation Report has to, "set out energy conservation measures that the authority considers practicable, cost-effective and likely to result in significant improvement in the energy efficiency of residential accommodation in its area." It will include assessment of costs of improvement measures, carbon dioxide emissions reductions, statements of local authority policies, effect on job creation, individual savings and so forth.

3.4.3 Although the Home Energy Conservation Act is a step forward, the down-side is that it is concerned only with residential accommodation. It does not concern itself with the commercial, industrial nor public administration sectors. It gives local authorities no powers to make grants or loans and makes no additional money available - except the Secretary of State's expenses in implementing the Act. At least every local authority is now, de facto, an "Energy Conservation Area", which should put energy conservation onto many more political and managerial agendas.

3.5 ENERGY LABELLING

3.5.1 When the 1981 Energy Conservation Act was in preparation some Dept. of Energy officials had hoped to be able to include provision for the mandatory labelling of energy-using appliances. This would enable prospective purchasers to make informed choices on the basis of comparative energy efficiency. In the event, their efforts were unsuccessful.

3.5.2 In 1984 the energy labelling issue re-emerged as a brain-child of two of the then newly appointed "Energy Marketeers", who had replaced the Advisory Council on Energy Conservation (ACEC), which had previously been the main source of expert advice on energy conservation for the Secretary of State for Energy. Their proposal, however, was not for the labelling of appliances but for a simple energy grading system for houses. As for previous appliances schemes, it was designed to enable potential purchasers to make informed choices. Their idea was based on the well established process by which motor-car manufacturers have to publish official Government fuel-consumption figures for new cars.

3.5.3 The idea came to nothing at the time but the incident illustrated the tensions between acknowledged 'experts' and the government, where positive action and legislation is concerned. The Government would only contemplate a voluntary scheme but their marketing advisers were saying in their Department of Energy Press Release:

> "in the interests of the public, Government should consider legislation if there has been no progress with a voluntary system within two years".

In a supporting statement they went so far as to state:

> "- by the year 1990 it should be illegal to own a property where the standard is worse than D (on their proposed rating system), ie. that of the current Building Regulations".

This was, indeed, strong stuff from advisers appointed by an administration keen on voluntary schemes and so pathologically opposed to coercive legislation.

3.5.4 There were five "Energy Marketeers" appointed in 1984 to advise the Secretary of State for Energy. Some of them are still valiantly fighting the energy conservation battle but, as a group, they gradually faded away, in spite of being heralded by the Secretary of State as, "...five very lively, able and experienced people to provide me and the nation with advice and ideas on how we can massively improve our energy efficiency in the coming years". On the other hand, he had also said, "I wish to make it clear that these five will not be a committee". He was probably hoping that they would not be as irritating to Government as the Advisory Committee that they replaced, which even the Secretary of State was on record as describing as, "hard hitting". Having killed the messenger council, the government soon discovered that the replacement messenger team was likely to deliver the same message.

3.5.5 By 1993 the issue of energy labelling of buildings was still alive, with the Advisory Committee on Business and the Environment (ACBE), like ACEC and the "Five Marketeers", calling for legislation. Again, the proposals fell on deaf ears. It is well acknowledged, by officials and professionals alike, that voluntary schemes in this area will not work. Regional Electricity companies have tried voluntary energy labelling schemes for electrical appliances but little progress has been made. British Gas was forced by its Regulator to display energy performance labels on new gas appliances in its showrooms - a sort of quasi, back-door legislative process - but even that has not led to universal labelling by all gas appliance retailers. As British Gas closes more showrooms, so fewer labelling opportunities will exist because the requirement was left to the Gas Regulator, rather than to general legislation, and OFGAS controls only British Gas showrooms.

3.6 BUILDING REGULATIONS

3.6.1 It had been said often that U.K. Building Regulations - as far as energy performance is concerned - put the U.K. standards for energy efficiency where most other E.U. countries were between the Wars. In 1994 new regulations were introduced to improve this situation and will come into force on 1st July 1995.

3.6.2 The 1994 Building Regulations are probably the most positive measure taken to improve the U.K.'s energy efficiency since 1979. The Building Research Establishment claims that the changes will improve energy performance of space and water heating in new houses by some 25%, with similar improvements for the non-domestic sector. However, this sort of figure is always subject to technical discussion, in which some might argue for figures as low as 12%. In any event, even somewhere between the highest and lowest, the lay-man can accept that the potential is , "a lot". What is also interesting, in view of the long and disappointing history of energy labelling, is that the revisions introduce the concept of compliance by conforming to an Energy Rating Method, based on the Government's Standard Assessment Procedure (SAP), which is, in effect, a buildings energy labelling system on a scale of 1-100.

3.6.3 The not-so-good-news about the improvements is that Building Regulations apply only to new-build. They do little to improve the existing stock of buildings, apart from a "trickle-down" effect on building practice in general. As new-build only affects a minute proportion of the stock, it has little short-term effect on the nation's energy consumption. However, a further innovation in the 1994 updating is that, for the first time, Part L of the Regulations, concerning energy conservation, will apply to certain building conversions - such as barn conversions.

Perhaps a greater disappointment was that proposals were dropped that would have severely constrained a building designer's freedom to include air-conditioning. As long ago as in the Seventies the Chairman of ACEC, Sir William Hawthorne, had been suggesting that most air-conditioning in the U.K. would be, "an energy obscenity".

It is also of interest that regulation now extends to space-heating and water-storage system controls, plus insulation of "vessels, pipes and ducts", which gets close to issues such as energy labelling of appliances, where legislation is so much needed.

3.7 FUEL SUPPLIER REGULATION

3.7.1 Depending upon one's interpretation of the term "efficient supply", it was possible to argue that, on being "nationalised", the publicly-owned fuel suppliers - coal, electricity and gas - had a statutory requirement to promote energy efficiency, even fuel conservation, among their consumers. In reality, those suppliers did not choose to place such an interpretation on their terms of reference and did not consider that they had any legal obligation that would involve positive action for energy efficiency. However, when duty called, they responded variously to Government requests for support in response to the oil crises of the Seventies; as did the oil suppliers - some commendably, others minimally. All four - oil, gas, coal and electricity - even went so far as to make co-operative efforts in support of Government initiatives but this, and their individual efforts, were entirely voluntary and remarkable for high levels of perceived activity with low levels of energy actually saved. Many government programmes to promote energy efficiency, both before and after the change of political flavour in 1979, were well supported by the fuel suppliers but the high levels of support gradually diminished as the fuel utilities were 'sold off' into the private sector. The oil suppliers had always been minor players, compared to the utilities.

3.7.2 As the fuel supply privatisation took effect there was speculation that far-reaching energy efficiency duties might be imposed on the energy industries, if only to restore the pre-privatisation levels of activity. In the event, the new duties were not particularly onerous and the leading player in energy conservation activity, British Gas, found itself spending less under the new requirements than it had as a publicly-owned body acting in voluntary concert with Government. Such expenditure continued to fall as the company felt its feet as a private monopoly operator. This was not surprising. Even after the knock-down prices that it fixed for the sale of energy utilities - which the late Harold Macmillan, had described as, "selling the family silver" - Government could not sensibly take measures to reduce overall fuel markets, that might deter would-be buyers.

3.7.3 Where the Electricity Supply Industry was concerned, the privatising legislation came to include a measure known as the Non-Fossil Fuel Obligation. It was presented as being imposed (at 10% of electricity end-user prices) to provide support for the production of energy from renewable sources - an admirable, environmentally friendly piece of legislation. In effect, the bulk of that money goes to supporting the costly and commercially unattractive nuclear electricity programme whilst, at the time of writing, little over 100 non-nuclear renewable projects have benefited. As energy efficiency legislation the measure was largely irrelevant.

In time the Industry's Regulator, OFFER, amended its rules to require Regional Electricity Companies to raise £1 per customer as a "allowable cost", to be used to finance about £100 million of expenditure over four years (from 1994) on energy efficiency. Like the gas Regulator, OFGAS, OFFER was more interested in reducing fuel prices than promoting fuel saving.

3.7.4 The privatised gas industry was subject to a different approach in its Authorisation Document. From 1986, when it was privatised, British Gas had certain duties concerning energy efficiency but these were confined to preparing statements, "setting out general information for the guidance of tariff customers in the efficient use of gas..." and, for contract (larger user) customers, the entirely non-prescriptive and non-onerous duty merely to tell them what energy efficiency services were available to them. There was no attempt to prescribe what level or type of advice was expected. There was certainly no guidance on how much should be spent, either in real terms or as percentage of profit or turn-over.

3.7.5 This situation was improved in 1991 when OFGAS devised and imposed on the Industry what became known as, "The 'E' Factor", ('E' for 'Efficiency'). By this arrangement British Gas was to raise money by an impost on the price of gas that would be used to finance energy efficiency projects. At a later date a different Regulator was to protest that, "It was not OFGAS' intention to require cross-subsidisation between tariff customers or uncommercial subsidisation of energy efficiency by the gas industry", but it was difficult to describe the 'E' factor as anything other than that, because it was used solely to finance subsidies. Described by some as, "Social engineering, by subsidy, financed by hypothecated taxation", it broke every dogma of the ruling Government!

3.7.6 The 'E' Factor was closely associated with the creation of the Energy Saving Trust. The gas-only, original body, the Energy Savings Trust, was to be the vehicle for administering the 'E' Factor funds (Part 6 contains the full story). Expenditure was to run into tens of millions of pounds. At the end of the first year a mere £100,000 had been spent, mostly on overheads, and only a cumulative £4 million or so by 1994. The whole edifice then became very unstable when an incoming new Regulator effectively put paid to the 'E' Factor concept.

3.7.7 In 1991, in parallel with the 'E' Factor venture, OFGAS imposed on British Gas a further piece of quasi-legislation to promote energy efficiency. It agreed what was termed, "A Statement of Intent", which produced a new Code of Practice - still theoretically voluntary - to stiffen the original requirements of the privatising 1986 Gas Act. The Code included better arrangements for staff training to give energy advice, for energy labelling of certain appliances and for the provision of information and advice, such as energy efficiency telephone-accessed advice desks.

3.7.8 In 1993 a further development replaced the voluntary Code of Practice with mandatory requirements. The Competition and Services (Utilities) Act was used to introduce the concept of "Standards of Performance" - an entirely misleading term, which should more properly have been "Operational Requirements" - which set down certain energy efficiency activities that had to be carried out. The same concept became applied to the Regional Electricity Companies.

3.7.9 The full details of what the standards of Performance require of the monopoly energy utilities in terms of energy efficiency are well documented by the Regulators and are continuously evolving. Similar arrangements do not apply to oil or solid-fuel suppliers but they do represent a form of energy efficiency legislation and provide proof that only legislation will get anything worthwhile done in this area. On the other hand, between the commercial acumen of the energy suppliers and the perversity of the Regulators, very little is being achieved in the way of fuel conservation and Government has effectively legislated itself out of having much, if any, power to do anything about it.

3.8 GENERAL LEGISLATION

3.8.1 Although there is growing acceptance of the role of legislation and regulation for energy efficiency and fuel conservation, it remains a truism that bad legislation can do untold harm to the cause it seeks to promote. From the "Save It" days, we still have legislation to control the use of energy to heat offices. It was so badly framed that it soon fell into disuse and gave unfortunately easy ammunition to those opposed to any effort to reduce energy use. In some cases, companies were found to be using energy to keep within the requirements, which was supposed to encourage people not to use it!

3.8.2 Of all forms of energy use the transport sector is the one that is most out of control and which is probably doing the most harm to personal health and to the environment in general. The motor car fuel consumption regulations, requiring manufacturers to state comparative consumption figures for new vehicles, were an excellent first-step but no further energy efficiency legislation has been forthcoming. 1995 saw emissions control legislation, focusing on the health and environmental problems and, indirectly, involving fuel consumption reductions among the offenders. It is to be hoped that this represents a change of heart by the Department of Transport that has been so reluctant to do anything about energy conservation in its sector of concern.

3.8.3 For the future, the energy conservation lobby can only hope that common-sense and the growing pressures from health and environmental protection and international carbon reduction commitments will influence official attitudes to re-admit legislation and regulation into the role of government. Perhaps the one most effective piece of low-cost-to-the-public-purse legislation would be to require all those who sell or let buildings to declare the structure's energy-rating. It has worked with new motor cars, has worked without setting potentially controversial norms and has not brought the industry to its knees.

In industry, legislation requiring significant energy users to employ certificated energy management, as is being pioneered in Japan, would also be an energy-saving measure that would have little cost for government. As with statutory energy-rating declaration for buildings, such legislation would stimulate existing energy efficiency businesses, create others and be able to operate very largely on existing legislation, precedence and professional practice. Enforcement would neither be particularly onerous nor costly; since compliance or avoidance would be clearly self-evident and, at least where buildings are concerned, transactions are already made within established professional bodies.

3.8.4 During meetings of the Advisory Committee for Business in the Environment (ACBE) and during the "Climate Change, Our National Programme for CO2 Emissions" deliberations, members of motor manufacturing and trading interests admitted that nothing had really progressed, as far as declaring fuel consumption was concerned, until the fuel consumption regulations had forced action. Voluntary schemes and self-regulation are notoriously ineffectual where major issues are concerned. Members made these admissions when referring to energy-using products other than their own motor vehicles. It is to be hoped that the example of the motor industry can be extended to buildings, energy-using appliances and industrial processes. Legislation will, ultimately, prove to be a necessary route to a low-energy society.

4 LOBBYISTS AND NON-GOVERNMENTAL ORGANISATIONS

4.1 The U.K. is well-served by non-governmental energy conservation organisations; for example, the National Energy Foundation, the Association for the Conservation of Energy, The Energy Savings Trust (See Section 6.) and Neighbourhood Energy Action (See Section 7.). These are the four major players but there are many more smaller, sometimes local, bodies; either seeking to raise energy conservation to a higher point on the political agenda or trying to gain advantages for their own members or client groups. They often work together but are sometimes in ideological conflict or just forced into conflict, competing for shares of scarce funding. Unhappily, expert and dedicated though they are, they have little financial strength and often themselves rely on external assistance; for example, Neighbourhood Energy Action (NEA) is about 50% financed by government departments, with most of its remaining costs met by private sector partnership or sponsorship. The Association for the Conservation of Energy (ACE) is 42% financed by the subscriptions of its industrial and commercial members - who contribute to establish their place in the energy conservation scene and to stimulate the market for energy conservation projects. Even so, this subscription funding has to be supplemented by further sponsorship for many of its most useful projects. Such bodies cannot provide finance nor resources for energy conservation initiatives but they can provide invaluable co-operation and guidance for projects that come already funded - or can lend experience and reputation in raising funds.

The four major players are sometimes referred to, either individually or collectively, as, " The Energy Efficiency Office in Exile", which says more about the plight of the EEO than about these valiant exiles. (Using the expression, " the plight of the EEO" is to comment on current political philosophy rather than on the Office itself and its Officials. They do much to encourage non-governmental energy conservation organisations with the very limited powers and resources allowed to them - to say nothing of further problems imposed on them by various non-elected and non-public-answerable quangos and Regulators).

4.2 THE NATIONAL ENERGY FOUNDATION (NEF)

4.2.1 The origins of the National Energy Foundation (NEF) go far back to the Energy Efficiency Unit of the Milton Keynes Development

Corporation, set up in the late Seventies but now defunct along with others of that sort. Early on in its existence the Corporation determined to make Milton Keynes a low-energy settlement and later moved to setting stringent standards of energy efficiency that it required of any building developer operating in the New Town. Such standards went far beyond the U.K. building regulations and were the basis of the National Homes Energy Rating Scheme (NHERS). The Corporation had also organised energy conservation exhibitions, including "Energy World" in 1986, which demonstrated that low-energy building is entirely compatible with design excellence and the practicalities of living.

In the early Eighties the Corporation sought to establish an energy conservation education theme park at Milton Keynes, seeking funding from energy suppliers. It was a far-sighted and potentially valuable initiative but the idea, likely to cost around ten-million pounds, happened to coincide with recession and the organisational turmoils created by the privatisation plans for the prospective main funders, the energy utilities.

4.2.2 A weakness of the project was that the Milton Keynes Development Corporation planned to design and create the whole project (theme park, permanent trade exhibitions, energy conference facilities, information centre and so forth) and then to form a foundation to manage it and to carry it forward. Some of the potential funders pointed out to the Corporation that this was to put the cart before the horse. The appropriate people were unlikely to come forward to serve on such a body, if they were presented with such a comprehensive fait accompli. It would be more reasonable to seek a body of expert and informed people and then invite them to map out a plan for the initiative, which may or may not have involved the facilities assumed by the Corporation. In the event, this advice was taken and, because of the lack of funding for the original proposals, this turned out to be the most practical way forward. Although the theme park was not built, the National Energy Foundation was formed and came into being in 1990 with a strong and well-placed governing body. Besides the energy theme park proposals, the Milton Keynes Development Corporation had handed over to the Foundation some established and on-going energy conservation services, including the National Homes Energy Rating Scheme (NHERS), which is described in Section 9 and which has continued to prosper. NEF also carried forward the Energy Park low-energy settlement and the low-energy Business Park. Both have - and continue to produce - excellent examples of low-energy buildings. There is also an energy monitoring service that is gathering information on the behaviour of the buildings and the settlement as a whole.

4.2.3 The creation of the National Energy Foundation happened to coincide with diminishing official interest in energy conservation, which could explain why it was often referred to as part of, "the EEO in exile". At times, especially when the possibility of the establishment of an Energy Efficiency Agency was debated (a phenomenon of the Early Eighties among the energy conservation "chattering classes"), the NEF was looked upon as a possible basis for such a body. These thoughts came to nothing but the Foundation prospered, in spite of energy conservation becoming increasingly politically unfashionable. In 1992, when Government hastily put together its Energy Saving Trust, it was a matter of some surprise that the established NEF was not given the function. However, when the Energy Saving Trust's Local Energy Advice Centres (LEACs) initiative was launched, NEF took on the overall management and contracting of the scheme; so it had clearly been a matter of politics not competence. The Foundation's contact address is : 3, Benbow Court, Shenley Church End, Milton Keynes, MK5 6JG. Tel: 01908 501908. Fax: 01908 504848.

4.3 THE ASSOCIATION FOR THE CONSERVATION OF ENERGY (A.C.E.)

4.3.1 ACE is a lobby and research group that was formed in 1981 by a group of companies that were, and are, active within the energy conservation industry. These are those concerned with the production, distribution and installation of insulating and draught-proofing materials, controls manufacturers, energy efficiency consultants and even the wholly-owned energy conservation subsidiaries of several Regional Electricity Companies. Its management is known to be considering how to involve other energy supply companies, which have energy conservation subsidiaries. Membership is confined to 20, being the major players in each field, and is made-up at Chairman and Chief-Executive level. This ensures a high degree of representation and member commitment.

4.3.2 The Association's objectives are to encourage a positive national awareness of the need for and the benefits of energy conservation, to help to establish a sensible and consistent national policy and programme and to increase investment in all appropriate energy saving measures. It produces a modest but authoritative newsletter, "The Fifth Fuel", organises research programmes that produce excellent reports and, in 1993, produced a guide for what they termed, "Housing Professionals", which will probably rank as one of the key text-books in energy conservation. ("Energy Efficient Homes". ISBN 0 901607 77 1. £15 from ACE or the Institute of Housing).

4.3.3 ACE is often dismissed by government Officials as, "just a lobby group" but, even though it is a lobby group, the Word "just" is undeserved and probably reflects official irritation rather than a serious judgement. It obviously has to consider the interests of its members but it has also developed an excellent reputation for sound research. It produces a regular series of "ACE Briefing Notes" on topics of immediate interest in energy conservation. It was a prime-mover behind Private Member attempts at an Energy Conservation Bill and was a sponsor of the lobbying for the Home Energy Conservation Act, which it successfully saw into law in 1995. The Association is constantly sought by the media for authoritative comment, having a record of 12 or more major articles a year in national newspapers.

4.3.4 One of the Association's past and present Chairman was a Member of ACBE and brought the considerable resources of the Association to that body's deliberations. It is represented on BSI and CEN committees and has a good record of giving evidence to the House of Commons and House of Lords Select Committees. Its Director, was, for a time, a Special Advisor to the House of Commons Environmental Committee.

4.3.5 ACE has recently spread its influence directly into Europe with its creation of EuroACE, which is based in Belgium to lobby European institutions. It has also launched an associate membership, Friends of the Association for the Conservation of Energy (FACE), for local authorities and housing associations, directly linked with the implementation of the 1995 Home Energy Conservation Act. The Association's contact address is 9, Sherlock Mews, London, W1M 3RH. Telephone: 0171 935 1495. Fax: 0171 935 8346.

4.4 THE INSTITUTE OF ENERGY

4.4.1 The Institute of Energy, formerly the Institute of Fuel (1979) and, before that, the result of a merger between the Institution of Fuel Economy Engineers and the Institution of Fuel Technology (1927), has had and continues to have a strong place in energy conservation. As an academically respected institution, it is not surprising that it has taken a leading part in promoting energy efficiency in education at all levels. What is more surprising is that it does not quite live up to expectations in this area. Perhaps that is because its Members are drawn chiefly from the fuel-supply industries and the emphasis of its educational initiatives is on the energy situation as a whole, rather than on efficient use and fuel conservation. In the early Eighties one of its Members, Bryan Smith, addressing a national Institute of Energy meeting, coined the famous dictum: "Not, 'Do not use' but 'Use Wisely!'", which did much to explain the Institute's intellectual situation in energy conservation issues.

4.4.2 The Institute's direct energy conservation efforts and initiatives can often be traced to a small group of members who are both experts and enthusiasts in the same area and who have been able to bring the majority of the membership along with them. This is not an uncommon situation within the generality of professional bodies, especially where a profession's members may have a business interest in the use of fuel in itself or in the production of fuel-using structures and installations. Architecture and other building professions are examples but, even in the CBI, the energy conservation body was only a sub-committee of a powerful, energy-supply-led, Energy Committee and has recently ceased to be active.

4.4.3 Although the Institute of Energy does make many excellent and exemplary contributions to the promotion of energy efficiency, many of its members are still of the "Institute of Fuel" inclination. However, it must be remembered that it was largely responsible for keeping CREATE alive and its energy conservation enthusiasts are powerful contributors to the whole scenario of a low-energy society.

The Institute's contact address is: 18, Devonshire Street, London, W1N 2AU. Tel: 0171 5807124. Fax:0171 580 4420.

4.5 THE ENERGY SYSTEMS TRADE ASSOCIATION (ESTA)

4.5.1 The Energy Systems Trade Association (ESTA) was established in 1982 to represent the interests of reputable suppliers of energy management products and services. Part of this activity is to promote the financial and environmental benefits of energy efficiency to energy users. Since its foundation it has prospered and grown and by 1995 had over 85 members from the leading energy management companies.

In pursuing its aim of representing only "reputable" companies, ESTA includes among its membership eligibility criteria that companies have to have been in business for at least 2 years and have to provide recent customer references. Its internal interest groupings also indicate the business scope of its membership:

- Contract Energy Management Group (CEMG).
- Building Controls Group (BCG).
- Independent Energy Consultants Group (IECG).

- Metering and Monitoring Group (MMG).
- Local Energy Controls Group (LECG).
- Control Systems Specialist Group (CSSG).

4.5.2 The Association has become sufficiently well established and respected to be represented on many BSI and CEN technical committees and has been a point of consultation for the Energy Select Committee and the energy control aspects of Building Regulations. It has strong overseas contacts, including representation on the European Federation of Energy Management (EFEM) and on the CADET Liaison Group of the International Energy Agency (IEA) of the Organisation for Economic Co-operation and Development (OECD), which has an Energy Policy Committee.

4.5.3 ESTA arranges national and regional seminars and exhibitions and organises NEMEX, the annual conference and exhibition of energy conservation interests, which was formerly run by the Energy Efficiency Office. Perhaps the Association's most important contribution has been its Energy Efficiency Accreditation Scheme, which it developed with the Institute of Energy. This scheme is detailed elsewhere (Section 11). Although the EEO put some funding into the venture, it is remarkable that ESTA was able to make it a success against a background a neglect and even hostility from sectors that one would have expected to be sources of support. That the scheme is so un-British as to include rigorous criteria and examination may explain why it had such a difficult genesis.

4.5.4 Of all the independent energy efficiency organisations, perhaps ESTA is one of those that most deserves the description, "The EEO in Exile". It has taken a strong lead, mustered reputable leading businesses and professionals and has an impressive record of achievement. The ESTA Energy Efficiency Year Book has an excellent directory of Members, which itself is a good introduction to this area of energy conservation, plus useful articles and case studies.

The Association's contact address is: PO Box 16, Stroud, Gloucestershire, GL6 9YB. Tel:01453 886776. Fax:01453 885226.

4.6 THE WATT COMMITTEE

4.6.1 In the late Seventies and early Eighties the Watt Committee on Energy was an important player in energy conservation matters and had established what one of the Government's Chief Scientists, Dr. Marshall, (1976) was describing as, "a continuous dialogue between the Watt

Committee and Government." Although it has since waned in influence and activity, its many publications remain key textbooks of many aspects of energy conservation, both policy and technicalities.

4.6.2 The organisation was set up in 1976, chiefly on the initiative of the Institute of Mechanical Engineers, to provide a single, independent forum in which professional people could, "contribute by all possible means to the formulation of national energy policies". Although this aim covered the whole range of energy policy - both supply and usage, the Committee has been predominantly concerned with energy efficiency and fuel conservation. However, like the Institute of Energy, which is a Watt Committee Member, and so many other bodies, its most recent emphasis has been with supply; for example, during 1994/95 it organised a series of seminars under the title, "Energy Now and the Next 50 Years". The four seminar subjects were,:

- Energy Resources and Scenarios.
- Energy Markets.
- Energy Policy.
- Energy Now and Over the Next 50 Years.

Efficiency and conservation were discussed but the overall titles are indicative of the way things have changed on the U.K. energy conservation scene, leaving the demand side still a poor relation to the supply side.

4.6.3 At about the time that the 94/95 seminars were running, the Committee was also leading a project, "Domestic Energy and Affordable Warmth", which was a new departure into a field not previously examined by them and a welcome re-examination of the Fuel Poverty scene by groups other than those traditionally associated with the subject; that is to say, those that are closely involved. The outcome was Watt Committee Report Number 30, "Domestic Energy and Affordable Warmth"; as with other reports of the Committee, something of a text-book on the subject.

4.6.4 The Committee's membership is drawn from 45 institutes and professional associations, plus about two dozen Associate Members, drawn from individual companies and organisations. Its principle sponsors are drawn from energy suppliers and some banks.

As said above, the Committee's position has diminished in recent years but chiefly because of funding restrictions and a general reduction in the amount of time and effort that individuals are able to devote to its activities. This has been a characteristic of energy conservation groups in most individual professional bodies and also of interest groups in general; perhaps merely a symptom of declining industrial and commercial activity and expertise and a cost-to-the-nation of the vogue for "de-layering" and "down-sizing".

The Watt Committee's contact address is; Burlington House, Piccadilly, London. W1V 0LQ. Tel: 0171-4343988. Fax:0171-434-3989.

4.7 THE COMBINED HEAT AND POWER ASSOCIATION (CHPA)

4.7.1 It is sometimes difficult to position the Combined Heat and Power Association (CHPA) between the supply-side and the demand-side of energy conservation but, on balance, it is more of an energy efficiency concept than an energy supply system. Whatever its conceptual position, it has established itself as a major player on the energy conservation scene and was chosen to manage one of first Energy Saving Trust schemes.

4.7.2 The organisation goes back to 1967, when it was known as the District Heating Association. It changed its name in 1983 to the Combined Heat and Power Association and was re-launched in 1988 with full-time professional staff. By 1994 it had a five-regional branch structure: London, Midlands, North West, Scotland and Yorkshire/Tyne-Tees besides its London Head Office.

4.7.3 It defines its position with a strap-line to its logo, "Energy Efficiency for a Better Environment". CHP is the process of producing electricity whilst recovering and putting to use all the waste heat that is generally thrown away. Conventional electricity generation is little more than 34% thermally efficient. With Combined Heat and Power, the technology exists to achieve up to 90% efficiency. The CPH Association exists to promote the wider use of CHP. It has over 100 members, including local authorities, equipment suppliers, consultants, energy management companies, electricity and gas suppliers, plus users of both CHP and community/district heating. It is important to register that CHP is not merely an adjuct of community heating but is also used in medium- and small-scale applications in both industry and commerce, in situations where there are substantial and constant needs for power and space or water heating.

4.7.4 Besides its yearbook, which gives details of member companies and exemplar projects, the CHPA has two regular newsletters. It is also a member of COGEN Europe, an organisation created to promote the wider use of CHP across Europe.

4.7.5 CHP itself has strong backing from the U.K. Government, which has set a target to more than double the U.K. use of CHP by the turn of the Century, and is supported by all the main U.K. political parties, the Confederation of British Industry, the TUC and many environmental groups. It is a curious twist of energy conservation politics that the Gas Regulator has gone so far as to make a strong endorsing statement in the Association's 1995 Newsletter but has also effectively scuppered the Energy Saving Trust's CHP Scheme (Managed by CHPA), by stopping its 'E' Factor funding!

The contact address for the CHPA is : 35/37, Grosvenor Gardens, London SW1W 0BS. Tel: 0171 828 4077. Fax: 0171 828 0310.

4.8 THE CONSTRUCTION AND BUILDING DESIGN PROFESSIONALS

4.8.1 For many decades the energy conservation lead among the construction and building design professionals was taken by the Royal Institute of British Architects (RIBA). In 1978 they obtained joint-funding from the electricity and gas utilities, from British Coal and from the Petroleum Industries Association to establish and staff the Energy Conservation Co-ordinating Group. Its main objective was to improve the energy conservation literacy and numeracy of the profession by arranging basic and mid-career education, exhibitions and conferences and by promoting the use of thermal evaluation programmes and other low-energy design tools.

4.8.2 It later changed its name to the RIBA Energy Group and enjoyed strong support from the RIBA Council and successive Presidents. It was remarkable for bringing together the "Four Fuels" in co-operative energy conservation ventures and for establishing close co-operation between its own academics and practitioners. (The fuels, of course, at that time, were only too pleased to have a low-cost, low-effort opportunity to make a high profile gesture towards supporting energy conservation. As official commitment was visibly reduced and as the fuel privatisation processes proceeded, so coal, electricity and gas gradually lost their enthusiasm for the Energy Group. Strangely, it was the already private oil industry that first pulled out completely, which might have warned politicians of one of the societal costs of "privatisation").

4.8.3 Whilst the RIBA Energy Group was developing, there was a parallel move by RIBA and other professionals to make a more co-operative grouping. This became known as "4-PEG" (The Four Professions Energy Group), comprising RIBA, CIBSE (Chartered Institute of Building Services Engineers), RICS (Royal Institute of Chartered Surveyors) and CIOB (Chartered Institute of Builders). Most of its activists were also involved with the RIBA Energy Group. The situation was not entirely clear because, by 1980, the RIBA Group had established a regional structure but based on already existing Four Professions groupings. The Architects had long considered themselves as the leaders among the many building professionals (RIBA = "Remember, I'm the Bloody Architect"!), so their Energy Group naturally sought to lead the Regional Energy Groups. A fairly regular newsletter was established, plus national working seminars, generally hosted and funded by one of the sponsoring fuel suppliers.

4.8.4 After a period of intense and useful activity in the Early Eighties, the RIBA Energy Group's fortunes waned. Whether waning interest caused the fuels to reduce and finally withdraw funding or whether the reduced funding was the prime cause of decline will remain a matter of continuing discussion. In any event, all four sponsors seemed pleased to have an excuse to withdraw. This co-incided with the long-overdue formation of the Construction Industry Council (CIC) in 1988, which drew together most of the dozen or so building-related professions and associations. Shortly after, this Council formed its own CIC Energy and Environment Group. Funding for the Group's secretariat was initially provided by British Coal, British Gas and the Electricity Association. Coal withdrew from this joint effort, followed by Gas in 1994. At the time of writing, only the Electricity Association remains, giving some cash support and leaving the CIC itself to make up the balance. It is greatly to the Construction Industry Council's credit that it has continued to staff the Energy Group. The Council represents some 330,000 construction industry professionals, besides other membership, and is well placed to further the cause of energy conservation and low-energy design in the U.K.

4.8.5 A less conspicuous, but important, construction industry body is the Building Energy Efficiency Confederation (BEEC). For the past 10 years or so it has arranged formal quarterly meetings with the Department of Energy - and now with, the Department of the Environment - chaired at Civil Service Grade 5 level with mutually agreed minutes. Its membership is drawn from all the trade associations in building. The contact address is: BEEC., 9, Sherlock Mews, London, W1M 3RH. Tel: 0171 935 1495.

The RIBA Environment Energy Group may be contacted through RIBA at 66, Portland Place, London, W1N 4AD. Tel: 0171 580 5533 Fax: 0171 255 1541

The CIC Energy Group's contact address is: Construction Industry Council, Store Street, London WC1E 7BT. Tel: 0171 637 8692. Fax: 0171 580 6140.

4.9 HOUSEHOLDER INITIATIVES

4.9.1 The special energy conservation needs of those on low-incomes are covered in Section 7, "Fuel Poverty", which is concerned with energy conservation among the "Fuel Poor". In contrast to that term the Energy Saving Trust has coined the term, "The Fuel Rich", to describe normal householders who can afford the fuel they buy, could afford to invest in energy-saving measures and who wish or ought to be doing something to reduce their energy consumption.

Well established groups, with wider remits, address some of their efforts to the energy conservation interests of the Fuel Rich; for example, Friends of the Earth, Greenpeace, the Consumers Association and so forth but, in recent years, two new organisations have entered the field under an environmental protection banner. These are ECOfeedback and Global Action Plan (GAP), who are becoming major players in this area.

4.9.2 ECOFEEDBACK

The Ecofeedback movement was set up in the U.K. in October 1993 but it had become established in the Netherlands since 1980, where it was devised by a physicist, Jan Hanhart, who recognised the power of feedback to promote energy user action from studying quality control in Japan.

The Netherlands scheme started with individual householders, was adopted by municipality energy utilities with Central Government backing and was finally taken over, as a nation-wide scheme, by the Netherlands association of energy utilities. When it came to the U.K., it was launched on a local authority basis and started with 12 authorities in 1993. In some localities it goes under the name, "Ecosave Campaign" and there are variations on the original Netherlands concept.

The Ecofeedback idea is that energy users will respond to information, which is presented at three levels:

- Present Performance: Householders record their energy consumption in a form that enables them to see how much energy they are using, to recognise trends and to assess their saving performance.
- Desired Performance: Local newspapers provide weekly reference figures, showing how much gas and electricity a given type of household should have used, based on type of building and weather, if it was energy efficient. Households then try to do better.
- Improved Performance: Local newspapers describe energy-saving measures.

Although Ecofeedback is mainly used for energy saving, the concept can also be used to monitor waste, recycling, transport energy and so forth.

In the Netherlands about 25% of households now participate and research has indicated a 10% saving by at least 60% of participants.

The contact address for information on Ecofeedback is: The New Economics Foundation, 1st Floor, Vine Court, 112, Whitechapel Road, London, E1 1JE. Tel: 0171 377 5696., Fax: 0171 377 5720.

4.9.3 GLOBAL ACTION PLAN (GAP)

Like Ecofeedback, Global Action Plan (GAP) also has overseas origins but is wider in scope and is more "environmental" in focus than purely energy conservation. The Plan was already running in the USA, Canada, Netherlands, Sweden, Germany, Finland and Switzerland when a pilot programme was launched in the U.K. in 1994.

Individuals may join the plan but it is more usual for teams to be formed; either of individual householders in a locality or based on the work-place, clubs or even pubs. "Ecoteams" receive Action Packs, usually meet once a month and, as with Ecofeedback, receive feedback to let them know the results of their actions - both on their home and on the country as a whole. Local authorities and companies may be involved, some in a cash sponsoring mode. "London First" (The curiously private-sector-devised "Banquo's Ghost" of the Greater London Council) has also backed the Plan in the Capital, along with a number of London Boroughs. The Plan targets savings on household energy, transport energy, water and recyclable waste. There is a quarterly newsletter, "GAP News".

The contact address for Global Action Plan is: 42, Kingsway, London, WC2B 6EX. Tel: 0171 404 0837.

4.10 This Section has not been an exhaustive coverage of the energy conservation lobbyists and activists in the U.K. but has high-lighted the few that have been, are or look set to be the major players. In addition there are many research organisations that have the subject as a major or single-issue, are variously funded and whose names crop-up across the energy conservation scene. Some, such as the Policy Studies Institute (PSI) or the Institute for Public Policy Research have a wider remit of which energy conservation is merely a part. Others, such as Energy Inform or Optima Energy, also provide consultancy and training services. Yet others, such as the Bristol Energy Centre (BEC) and the Energy Conservation and Solar Centre (ECSC) extend their activities to ventures such as running LEACs for the Energy Saving Trust. Of all these variations, some are private, profit-distributing others are "not-for-profit" organisations. Additionally, there are hundreds of individual energy efficiency consultants and, at a further extreme, the energy efficiency departments of very large consultancies; such as P.A. Consultants.

4.11 All major political parties in the U.K. have energy conservation special interest groups and there are parliamentary groups; such as PAEG (Parliamentary Alternative Energy Group), whose name belies its wider interest in energy conservation. Many universities have specialist energy conservation sections and the academics concerned often appear on the boards or among the membership of lobby organisations and advising government Select Committees.

5 THE ENERGY CONSERVATION PRESS

There is not much press specifically devoted to energy conservation; though some organisations produce useful newsletters and occasional publications. These range from the U.K.'s CHPA to the EU's "COGEN Europe", and others. Various professional bodies also use their magazines to cover the subject by publishing "Environment" or "Energy Efficiency" special issues or supplements. To date, these have generally been excellent and useful.

5.1 "Energy Management" was started by Bernard Ingham, in 1978, when he was head of the then Energy Conservation Division of the Department of Energy; before he went on to more notorious things in the Cabinet Office. It started life as a purely Government journal but has since been partly "privatised". Although it is now published by a private company, (John Ball), it retains an Editorial Board whose Chairman is one of the Directors of the E.E.O.. This Board includes representation from other DOE Directorates, D.T.I., ETSU and BRECSU. The publication still calls itself, "The Journal of the EEO". It gives wide coverage of energy conservation affairs, including Government policy and ministerial pronouncements, plus excellent technical and management features and advertisements. As an ex-journalist, it was probably not surprising that Bernard Ingham gave "Energy Management" to the U.K. energy conservation scene. In its early days it was very much a newspaper in style and content but has since moved to looking more like a very respectable journal of a learned profession. Like so many other initiatives, it represents something of a lost opportunity, finally falling victim to "privatisation mania", lacking the best of either the official or the private. It swung uneasily between the industrial/energy manager and the domestic/general public constituency when there was clearly a case for two publications. "Home Efficiency" (see below) may be filling the domestic gap but it can only be a matter of regret that there has not been an "official" publication for this important sector. Laudable private initiatives to fill the gap can never make up for the lack of a positive, official signal from government. On the other hand, it is all to easy for an "official" journal to stick too solidly to a government's "line". At present some energy conservation lobbyists refer to "Energy Management" as "PRAVDA".

Appropriate persons may receive the journal free on application to the EEO or the publishers: The Circulation Manager, "Energy Management", Maclean Hunten House, Chalk Lane, Cockfosters Road, Barnet EN4 0BR.

5.2 "Energy in Buildings and Industry" is a wholly private venture and the survivor of a number of past endeavours in this sector. Besides excellent technical and energy management features it has independent comment on energy conservation politics. In fact, of all the representatives of the Energy Conservation Press, "Energy in Buildings and Industry" provides the few crumbs of journalism that raise the issue above the pure technicalities of the subject. At the time of writing it still maintains a long-running column by the Director of the Association for the Conservation of Energy; a past winner of the Energy Journalist of the Year Award and a well-informed and measured critic of official policy. Like "Energy Management", it is distributed free to appropriate persons. Those wanting to be included in the circulation may write to PO Box 21, Westerham, Kent TN16 1BR. It lives up well to its strap-line, "Promoting Energy Efficiency".

5.3 "Energy Action" is the journal of Neighbourhood Energy Action (NEA) and chiefly concerned with energy conservation among low-income households. However, it does cover wider issues in its, "My View" feature which carries social and political comment by leading-lights in the energy conservation scene. Although it remains mostly related to the Fuel Poverty area, overall it offers perhaps the most comprehensive briefing available to those interested in energy conservation in the domestic area. It appears four times a year, free to NEA Members, and £15 a year by subscription. (Editor: Niel Ritchie, Neighbourhood Energy Action, 90-92, Pilgrim Street, Newcastle Upon Tyne, NE1 6SG.) If there is "essential reading" for studies on energy conservation politics and general affairs, "Energy Action" is it.

5.4 The Association for the Conservation of Energy (ACE) issues a regular publication, "The Fifth Fuel", which provides comment across the whole sector. It is available to ACE Members but it is also widely circulated on request. Although it is very useful, its information is not immediate - merely because of publication schedule - but it is among the most authoritative. The Editor is at ACE, 9, Sherlock Mews, London W1M 3RH.

5.5 In the same sort of category as the ACE publication is "The Energy Observer". This is a usually quarterly publication produced by the energy consultancy firm, AHS, Emstar Plc., giving useful comment on the energy conservation scene plus best practice case-studies from the Company's own client experiences. It has limited topicality, or immediacy of topicality, but is, nevertheless, a useful part of the overall U.K. energy conservation press. It is distributed free, on application to AHS, Plc., 50/60, London Road, Stains, Middlesex, TW18 1BR.

5.6 The Energy Systems trade Association (ESTA) produces its, "Energy Efficiency Year Book", which offers a good guide to the energy efficiency industry, not least a directory of ESTA members. It explains what each Member is doing or producing, plus providing editorial comment and "Good Practice" case studies. Even its copious advertisements give a useful snap-shot of industrial and commercial activity in energy efficiency. One can apply for inclusion in the circulation list to, ESTA, PO Box 16, Stroud, Gloucestershire, GL6 9YB.

5.7 With a wider remit than energy conservation, there is the magazine "Resource", launched in 1993, with the valiant policy of surviving without trade advertisements. It is aimed at those concerned with the management of energy and water resources but chiefly with energy management. It appears bi-monthly (6 copies plus 4 quarterly guides a year) with an annual subscription rate of £150. Its address is, Marketing Dept., Eclipse Group, Ltd., 18/20, Highbury Place, London, N5 1QP. Telephone: 0171-354-5858. Fax: 0171-354-8106.

5.8 "Green Energy Matters", concerned with the wider environmental protection scene, as its name implies, also appears bi-monthly and is mainly concerned with the politics of the issues - as opposed to the more practical management approach of "Resource". ("Green Energy Matters", editorial office is at PO Box 21, Westerham, Kent, TN16 1BR.) However, each, in its sphere, is well produced and represents a responsible journalistic treatment, though not yet enjoying a sufficiently wide circulation to represent an influential or authoritative voice.

5.9 Friends of the Earth Scotland has its own, "Safe Energy", which covers chiefly nuclear and renewable energy matters but with a significant energy efficiency content as well. It was published bi-monthly from its first appearance in 1976 but is now a quarterly publication (Subscription rate: £38 a year for organisations, £15 for individuals, for 4 issues). The editorial office is at 72, Newhaven Road, Edinburgh, EH6 5QG. Tel: 0131 5549977.

5.10 "Energy Report" appeared briefly around 1990/91 as the journal of the National Energy Efficiency Foundation but petered out with the

Association itself. The organisers hope to resume publication if the Association manages to revive.

5.11 A newcomer to the scene, towards the end of 1994, has been "Home Efficiency", a freebie which declares itself to be, "aimed at raising people's awareness of this important subject". It skilfully fills a gap in the matter of energy advice to householders, being a mixture of an "Energy Which?" and an Energy Efficiency Office booklet. Its constituency is not merely the bewildered householder, wondering what to do for the best, but includes those with an interest in the well-being of householders; such as, local authorities, housing associations and building societies - all anxious to secure their economic interest by ensuring that their tenants/borrowers do not waste money on wasted energy. 30,000 copies are distributed on a bi-monthly basis and the publication has the endorsement of the Energy Saving Trust, the National Energy Foundation, the National Housebuilders Confederation and the National Home Improvement Council. However, this is an "endorsement" not sponsorship or direct control. Nevertheless, it represents an important indication of status in the energy conservation market. It is published by the Efficient Books Company, Ltd., Hilton House, Grove Lane, Chalfont St. Peter, SL9 9JU - Tel: 01753 884216, Fax: 01753 885906.

5.12 Beyond the trade and specialist press, energy conservation media coverage is increasing and putting up a good fight for public interest against other social issues, hopefully to improve the level of attention that it occupies on public and official agendas. The excellence of what remain of the specialist publications reflects the enthusiasm and competence of those continuing to promote energy efficiency and fuel conservation. The widening of the scope of the energy management press, into environment and water resources, will be a useful addition in ensuring continuity and survival. However, even with an energy efficiency article appearing about once every 4 days in the national press (ACE research figure) general media practitioners have never had more than a tenuous grasp of the basic issues of energy conservation, have tended to gravitate towards exotic aspects and have been easily seduced by the "windmills and solar panels" - quasi-supply-side topics; much as many otherwise well-intentioned teachers are. The advent of "the Environment" and "Global Warming" will offer more scope for media attention but probably only at the expense of offering even more exotic and irrelevant distractions: "More noise in the system", as the media people themselves would say.

6 THE ENERGY SAVING TRUST

6.1 AN EXERCISE IN ENERGY CONSERVATION POLITICS

As an exercise in understanding the politics of energy conservation in the U.K. , the Energy Saving Trust (EST) must rank as one of the most valuable lessons in complexity, conflict, confusion and political manipulation, lost opportunity and wasted resources; plus the sad lesson that many otherwise valuable initiatives are only redeemed by the dedication and hard work of those employed to run them.

6.2 EARLY HISTORY

6.2.1 As far as the energy conservation public were concerned, the EST came to light in the autumn of 1991, when the Director General of Ofgas, Sir James McKinnon, speaking at the NEA Annual Conference, let it be known that such a body was likely to be formed. It had not been the subject of a formal press release. The more general public heard of it in the Spring of 1992, when it appeared in the Tory Party election manifesto. In fact, the concept evolved as far back as 1989, was put on a shelf and then taken down and dusted as a practical way of working the Gas Industry's 'E' factor, which was devised towards the end of 1990

6.2.2 During 1987/88 Neighbourhood Energy Action (NEA) was having discussions with British Gas about the possibility of providing very low-cost gas heating appliances for the "Fuel Poor". Good progress had been made, to the extent that some gas-fire manufacturers had proposals for producing basic, high-efficiency appliances at very low cost. Part of the cost saving was associated with the absence of any marketing and merchandising overheads, both in wholesale and retail terms. It was also intended that old casing designs should be used, for which tooling was available, stored and long written off, to carry up-to-date high efficiency mechanisms. As negotiations between British Gas and NEA proceeded, it also became evident, from a study by the energy researcher, Gill Owen, that, for the types of household circumstances likely to be involved, the running costs of a central heating system would be lower than the costs of using individual gas heaters. This also agreed with some British Gas studies. At the time (1988) calculations had suggested that a package could be produced giving installation re-payment costs of £5 per week, plus running costs of about the same sum. The proposed scheme had progressed beyond low-cost to "free" in some circumstances but there were problems arising from sales staff commissions and British Gas' regional accounting.

6.2.3 It was from these problems that the planning team devised the concept of a Trust to handle the funding of the scheme. In outline, but in reality far more complex, the scheme envisaged British Gas allocating a sum of money to pay for the installations outside its normal accounting system. The sum would be lodged in some sort of ring-fenced trust and used to meet accounts for the works, which would be raised and submitted by regional Sales Departments ; as though the trust body were, say, a local authority or housing association. The bill would be sent to that body instead of to the low-income customer. Qualification for the benefit was to have been handled through what were then members of N.E.A. (This was before the current EAGA and Network Installer arrangements and the means-tested benefit qualification of the Home Energy Efficiency Scheme).

6.2.4 The reasons why the scheme was put "on the shelf" are internal to British Gas - at the time related to the changes brought about by "privatisation". However, the existence of such a scheme was very convenient when the Gas Industry 'E' Factor came into being. Its fund-holding trust element provided a well thought through model concept for the Energy Saving Trust.

6.3 THE 'E' FACTOR AND THE GAS TRUST

6.3.1 The Gas Industry 'E' Factor has been well documented and discussed and is more concerned with the wider politics of energy supply; as opposed to those of energy conservation. For the purposes of this study it is sufficient to say that it was devised as a way to raise money from energy users to finance energy conservation. (The 'E' stands for "efficiency".). At first the sums involved were trivial, in terms of their effect on the price of gas ; two million pounds in the first year involved only a fraction of a percentage point per cubic metre of gas. However, it was envisaged that this sum would rise rapidly to tens of millions a year. (The unfortunate out-turn of these early plans is discussed below. At the time it seemed that a simple means had at last been devised to raise money for energy conservation). However, there arose a matter of audit and trust. Could British Gas be trusted to use the money exclusively for energy conservation purposes? Could British Gas trust the Energy Efficiency Office to use the funds effectively? It was then that the idea of a trust, as an independent body, was put forward by British Gas, based on the earlier work with Neighbourhood Energy Action. It was accepted by the Regulator (then James McKinnon), in the autumn of 1991, and work began to put it in place. Simultaneously work began to put together energy efficiency schemes, to be funded and directed by the trust when it would come into being. The resulting body was called the Energy Savings Trust. It envisaged a controlling board, a chief-executive and a small administration staff - most of the work of running schemes to have been contracted out to expert bodies. One intention of British Gas was that the new, independent body would be free of the established staff and administrative overheads of the large organisation and capable of running in a "lean and mean" manner, so that funds did indeed go to energy conservation without losses to overheads.

6.3.2 To understand the later problems that arose and to define lessons for future policy development in this area, it is useful to rehearse the basically simple features of the original Energy Savings Trust (EST). (The later body dropped the 's' from "Savings" and became, "The Energy Saving Trust"). First, the energy saving schemes were to be financed exclusively by revenue raised from tariff gas users. There was always a possibility of involvement of other sources of finance but only as an adjunct to the "gas" nature of the initiative ; and such extention was not an important part of the early thinking. The policy and major decision making processes of the Trust were to be carried out by the Board, closely informed by senior management of British Gas and the Company's Energy Conservation Co-ordination Department. This would mean that the Chief Executive of the Trust would have a mainly managerial role, in an administrative sense, and would be an important but not a 'top management' appointment. As an indication, the salary scale envisaged for the post was what British Gas then termed H.M. 2/3 - a higher-management position about three levels below H.Q. Director (In 1991 about £35,000 p.a.). The Board was to comprise some British Gas senior management, including British Gas Board-level involvement (as then constituted, but not comparable to the present (1994) organisational levels), plus "outsiders" representing and knowledgeable about the energy conservation scene. External advice was taken at a very senior level on remuneration. There was a strong vein of opinion that appointments could be on an "expenses-only" basis but the consensus finally formed around a token level of remuneration of £5,000 pa. for a Chairman and £1,000 for Board Members - excluding British Gas representatives, who would receive no remuneration from the Trust. It was also determined that only modest premises would be required, perhaps in a London suburban location, with 3-monthly board meetings held at a central London location on British Gas premises, if required.

6.3.4 From this it will be seen that the architects of the original body were intent on creating a lean, low-cost organisation, that would be proof against growing into a bureaucratic monster or of creating a life of its own, that might develop into a dog-wagging tail! It was not to be a policy-making body, merely a managing and auditing agency.

6.3.5 Development proceeded on these lines, with the intention of having the first energy-saving schemes in place by the early summer of 1992. Events took a sharp turn, involving a subsequent long implementation delay, when the manifesto needs of the Tory Party coincided with internal manoeuvring within the Civil Service to induce the then Secretary of State for Energy, John Wakeham, to ask the Chairman of British Gas (Robert Evans) to halt the announcement of the gas-funded Energy Savings Trust and to discuss British Gas involvement in a wider body, which would include Government and the Electricity Supply Industry. This move was swift and dramatic, involving a Friday-afternoon change to a following-week event, coming as a surprise to all concerned. (It is also unclear precisely how well the Regulator was consulted, which, perhaps, explains the later difficulties that arose between him and the wider body that was set-up).

6.3.6 Arrangements were changed rapidly and all development put on "hold" so that the creation of this exciting "new" initiative could be included in the Tory manifesto, where there had been a prominent lack of environmental or energy conservation commitment. (The tragedy was - considering the harm done by this manoeuvre to what had been a really new ground-breaking idea - that "the Environment" and energy conservation were not really mainstream issues during that election and little would have been lost or gained by leaving the original idea to take its course).

6.4 THE ENERGY SAVING TRUST

6.4.1 Once the election was out of the way the manifesto item of the Energy Saving Trust (EST) had to be implemented. In effect there was a one-year delay in implementing even a semblance of any of the original British Gas proposals for action. There was an up-hill struggle to involve the Electricity Supply Industry and a real problem in finding a suitable Chairman and Board. The Government take-over of the idea involved considerable changes in approach and scale, which created weaknesses and opportunities for trouble that would soon reveal themselves. The idea was now politicised and burdened with all the problems of fierce inter-fuel competition, plus the ruffled feathers of both the Gas and Electricity

Regulators. The Trust had few friends, even before it finally got off the ground, in 1993, and moved into its premises in Buckingham Palace Road.

6.4.2 From the appointment of its first Chief Executive (Dr. Eoin Lees) in November 1992, the organisation of the new body moved out of the hands of British Gas and the Energy Efficiency Office and was taken over entirely by the Board of Directors. However, British Gas had two directors on the Board, as had the Regional Electricity Companies. The Secretary of State for the Environment could, but did not, exercise a right to representation of the Board. Along with the gas and electricity interests, the Secretary of State is one of the Company's guarantors; as is the Secretary of State for Scotland and the Secretary of State for Northern Ireland.

6.5 DEVELOPMENT OF THE TRUST

6.5.1 For those operating in the energy efficiency/conservation field it is important to grasp that the Energy Saving Trust (EST) is not some sort of surrogate or privatised Energy Efficiency Office. In its early days its Chairman, Lord Moore, was reported as saying that the Trust would ultimately, "subsume the functions of the E.E.O.", but there was little further talk of that sort once the Trust became established and the limitations of its activities and finances became apparent. It cannot, therefore, be regarded as a source of funding or even advice for energy conservation projects; unless these already have funding agreed from Government or the fuel suppliers. In which case, the Trust is available as a project manager and independent director. Its Chief Executive is on record as saying that the Trust will not grow into a vast bureaucracy and that it will contract out most of its needs for project management and expertise.

6.5.2 On paper, the Trust is an ambitious and impressive undertaking. Its Corporate Business Plan envisaged an annual expenditure of £175 million by 1996. This money will be spent on projects that target owner-occupiers, tenants of social housing, small businesses, schools and low-income energy users. Even this sum will grow, to enable the Trust to reach the target, given to it by the Government, to deliver an annual reduction of 2.5 million tones of carbon emissions by the end of the century. This is to meet the U.K. Government's 1992 Rio Earth Summit pledge, to return U.K. emissions of carbon-dioxide to 1990 levels by the year 2000. By 1994 the Trust had an establishment of about a dozen, with only modest plans to increase that number. This modest establishment is in line with the Board's resolve to contract out most of its work - a principle adopted by British Gas and Ofgas when the original body was conceived of.

6.5.3 At first the Trust was able to get off to a good start. Two Schemes were handed over to it, virtually ready to start, which had been worked up by British Gas, Ofgas and the Energy Efficiency Office. These were the Owner Occupier Condensing Boiler Scheme and the Residential CPH Scheme. (A third Scheme, concerning low-income households, was at an advanced stage but ran into problems with the Regulator and, at the end of the Trust's first year, was still on the drawing-board). These two Schemes concerned gas users only and were wholly funded by the 'E' Factor levy on the price of gas, paid by British Gas tariff customers. Both as pilots, to test the contracted-out method of operating, and as energy conservation initiatives, the two schemes were successful. However, there was a change of Regulator for the gas industry at the end of 1993 and the subsequent difficulties, which were well documented in the press, resulted in the closure of the Condensing Boiler Scheme after one year and the announcement of closure of the CHP Scheme after two years, (its original pilot period).

6.5.4 When the Trust was set up there were three members and guarantors: British Gas, the RECs and Government, who were all expected to produce funding for Schemes. On the whole, the Government failed to produce any real "new" money and chiefly confined itself to "in kind" support, through staff secondments. The other members were, understandably, not too happy about this; especially British Gas, whose cash and resources had been used to set-up the Trust, develop its first Schemes and whose customers would be footing the bill to run the Schemes. It should also be remembered that the whole concept had been a British Gas initiative, for which some public credit would have been expected. Even before the Government stepped in to "hijack" the idea, the Industry's Regulator had made a unilateral announcement about it, without warning British Gas, and had, himself, carried out his own P.R. hijack of it. All along, in their dealings with the Regulator, British Gas had been accustomed to this sort of cavalier behaviour but, nonetheless, when applied to the Energy Savings Trust initiative, it was irritating for the Company to see an excellent and original gesture taken from its control and credit. Worse still, its management had to watch whilst start-up was delayed, costs soared and their customers gained little benefit.

6.5.5 After the first two Gas Schemes had been launched, the Government appeared to be producing cash to fund a Scheme which would pilot energy advice centres. These later became known as

L.E.A.C.s (Local Energy Advice Centres) and, in line with the Trust's policy, the management and development was contracted out - to the National Energy Foundation (NEF). Twenty-nine Centres were set up, spread across the whole of England, Scotland and Wales. This move did not entirely satisfy the other Members because the Government had planned and budgeted for some initiative of this sort even before the Trust was expanded into its tripartite membership form. An attempt was also made to attract cash from the other members into the Centres but met with some resistance, except on a very local basis from some of the RECs. The pity is that setting up such a network of advice services could have been an entirely logical and appropriate joint-funding exercise; had it been approached in that light from the start, rather than manipulated as an attempt to suggest real Government commitment to an effective share in funding the Trust's work.

6.5.6 As far as Electricity Industry funding was concerned, matters were not as straight-forward as with gas. In the first place, the RECs were being cajoled into a structure on which they had not been consulted. Secondly, their Regulator was not favourably disposed towards an 'E' Factor for the Industry. Thirdly, post-privatisation, the Industry was fragmented, both vertically and horizontally, resulting in production and transmission being separated from local distribution and marketing and resulting in a fierce competitive situation; not only across the RECs but also between the RECs and the generating companies; the latter, to date, being totally uninvolved in the Trust and its funding.

6.5.7 It was not surprising, given this financial back-ground, that it proved difficult to get any electricity Schemes moving. It is to the credit of the Trust's Board and management that at least one minor Scheme was agreed, though supported by only eight of the twelve RECs. The Scheme itself was to promote the use of compact fluorescent lights (CFLs). In under two months, in 1993, about three-quarters of a million units were sold, as a result of the Scheme's subsidy, roughly the same number as were sold in the whole of 1992. It was also an excellent exercise in raising public awareness of the cash-saving, energy-saving and environmental benefits of CFLs.

6.6 THE INITIATIVE STALLED

6.6.1 At the time of writing, no further significant Schemes are likely to be launched. The Gas Industry Regulator has stopped the two gas Schemes and has called into question the whole concept of the 'E' Factor. The Electricity Industry Regulator remains opposed to an electricity 'E'

Factor and has merely allowed the RECs to raise the modest sum of £1 per customer for energy conservation work. Even this sum does not need to be channelled through the Energy Saving Trust; an important consideration when many REC managers can see no benefit, and merely unnecessary overheads, occurring from such involvement. The £1 itself is modest when compared to the 10% Non-Fossil Fuel Levy on electricity prices, which raised £1.2bn in 1992/93. As most of that enormous figure goes to the nuclear industry (part of the supply side of the supply/demand equation), one can feel little but despair. Ten per cent levied for demand-side projects and the nuclear industry left to the market, would totally alter the U.K. energy scene and release cash and resources from energy supply to be used on productive industrial projects and social services. The sums raised from a 10% energy conservation levy would put the Trust well on its way to turning the U.K. into a low-energy society, meeting the Nation's Rio commitments and giving its citizens a less polluted environment to live in. Such a move would also be compatible with the Government's alleged principle, that the polluter should pay. After all, it is gas, electricity, coal and oil users who pollute by using. The producers merely pollute in response to the users' demands.

6.6.2 One matter that the Trust has tried to move forward has been the long-standing problem of motivating the "Fuel Rich" to engage in energy saving activity. (The "Fuel Rich" is a term coined by the Trust to designate those able to pay their present fuel bills and able to invest in energy-saving measures - if they so chose; "Fuel Rich", as opposed to the more usual and emotive occupants of the "Fuel Poverty" sector). Even before the Trust had come into being, a sub-committee of ACBE had been considering a scheme that became known as "Homes 2000", to encourage the Fuel Rich to become active in energy efficiency investment. Little came of the work, chiefly because of the reluctance of the fuel supply industries to fund the initiative. The Trust gradually took over the idea and "Homes 2000" and a "Social Housing Scheme" were both promised to begin in the autumn of 1994 in the Government's, "Climate for Change" White Paper, launched in January of that year by the Prime Minister himself. Faced with the same lack of funding for the ideas as ACBE had been, the Trust was unable to progress the official promises.

6.6.3 In 1994 the Trust engaged an independent, "not-for-profit" organisation, "Projects in Partnership", which has funding from the Department of the Environment, to develop and test practical solutions to the Fuel Rich problem. To date, they have produced three reports, which analyse the situation and suggest ways forward. The Reports are an excellent analysis but the conclusions are entirely predictable and will give little comfort to those whose touching faith in the Holy Grail of the "Free Market" prevent them from acting. Projects in Partnership have found, as have so many before them, that, "...the major barriers to increased energy efficiency in the home lie in the interests between homeowners and the technical sector, the credibility of the message and the lack of incentives...". From this, it is not surprising that, "Financial incentives are needed to reinforce awareness raising campaigns..", which is what ACEC was saying in the Seventies and ACBE in the Nineties. Is there anyone listening out there?

6.6.4 Given the unhappy, politically-manipulated history of the Energy Saving Trust, it is not surprising that it is proving an expensive, frustrating disappointment for the Energy Conservation movement. It has become a tangled web of political intrigue, manipulated to such an extent that its original aim seems to be lost. It was conceived to enable gas users to become more energy efficient. It was snatched from its cradle to satisfy a barren step-mother and now finds itself the victim of the machinations of some sort of energy Family Support Agency trying to pin paternity and a duty of financial support on anyone - so long as it is not the kidnapping step-mother! Against such a situation one now wonders whether a properly funded Energy Efficiency Office might not do the job more effectively, more cheaply and with some semblance of democratic legitimacy. Alternatively, there is the French model of the Agence Pour la Maitrisse d'Energie et L'Environment (AMEE), a well constituted and well funded organ of government.

6.6.5 The student of the politics of energy conservation may feel that the prospects for energy conservation in the U.K. in the hands of the Trust are bleak, that the prospects for the Trust itself are even more bleak. In May 1994 the Chairman of the Trust, Lord Moore, was writing to the Secretary of State for the Environment, "I view the current situation as serious and in need of clarification..." ... and he should know! However, the Trust has been able to encourage or participate in a number of energy efficiency activities, to which it is able to lend its name, if not any funding, and has demonstrated its potential. Realisation of that potential will rely on the funding activities of the fuel supplier Regulators. Who controls them seems to be anyone's guess and, at the time of writing, they each appear to regard the promotion of energy conservation to be outside their duties or, at least, marginal to their main interests.

6.6.6 In February 1995 there were signs that Government had begun to face up to the reality that the Trust is not going to be effective as constituted. Having avoided as much direct funding as was decently possible, the Secretary of State for the Environment announced that he was to add the Trust to a list of bodies that may receive financial assistance from central government for environmental purposes. He is empowered to do that under the Environmental Protection Act of 1990. The Trust's Chief Executive has stated that, "the Trust is a very different organisation from the one which was originally set up". He says that it is no longer driven by meeting carbon saving targets but has a role in changing the energy services market, particularly where electricity and gas competition are concerned. At least he can plan for the future without worrying about paying wages and rent, which is the very least he deserves for having had to run with such a slippery ball from the start.

To return to the irritating metaphors of paragraph 6.6.4, the Secretary of State's announcement suggests that the "kidnapping step-mother" is going to contribute to the child's support but within an agenda that will get the money back form the original parents - through a Gas Bill and whatever follows for electricity.

6.6.7 If nothing else, the history and current situation of the Energy Saving Trust demonstrates that, however excellent the idea, very little can be achieved without adequate funding. Where energy conservation is concerned, "adequate" means very large sums. Industry, "the Market", will only contribute such funding if it can see an immediate and substantial commercial advantage. In the absence of industrial or commercial funding, the money has to be provided from public money - either local or national. Privatised, the former big players are no longer under the control of the public and their Regulators are clearly not going to force them to pay up. Any new players in energy supply are highly unlikely to contribute funds to cut the size of the energy-use cake when they exist, commercially, to make profits from taking slices of it. So far they have been conspicuously absent from EST activities or funding. Government has found that the Trust is a far more expensive way of doing things than direct action by existing official resources, with the Trust and its funding quite out of their control. It has certainly been an interesting exercise in energy conservation politics.

7 THE FUEL POVERTY SECTOR

7.1 FUEL POVERTY

7.1.1 Largely because of the scale of the problem and because of its politically high profile - as fuel costs rise and poverty increases - "Fuel Poverty" has become a major issue on the U.K. energy conservation scene. When government sees fuel-pricing as a tool for reaching its environmental protection commitments the politics of Fuel Poverty become even more interesting. Studies by the institute of Fiscal Studies show that the price elasticity of fuel has an average value of -0.3 (10% rise in price results in 3% drop in spending). However, this masks a range of -0.06 for the richest 20% of households and -0.6 for the poorest 20%. The poor suffer disproportionately from a price-driven fuel saving policy.

The term "Fuel Poverty" became current in the Eighties and has been well described by Dr. Brenda Boardman of the Oxford Centre for Environmental Change, whose book, "Fuel Poverty" did much to give a factual and intellectual definition of the problem.

7.1.2 In essence, "Fuel Poverty" is caused by energy inefficient homes inhabited by disadvantaged citizens, who do not have access to capital to make energy-saving improvements; such as, adequate insulation or the installation of efficient, economical heating systems. Over 30% of U.K. households are in this situation - and increasing.

The most influential player in the Fuel Poverty scene in the U.K. is the charitable organisation, Neighbourhood Energy Action (NEA), which operates nationally from its headquarters in Newcastle-upon-Tyne and enjoys the support and confidence of politicians right across the political spectrum.

7.2 NEIGHBOURHOOD ENERGY ACTION (NEA)

7.2.1 NEA was launched in May 1981 as a development programme of the National Council for Voluntary Organisations (NCVO) but it can trace its existence to an initiative started several years earlier in Durham, where its founders had hit upon the bright idea of combining the power of government home insulation grants with job creation initiatives. In 1985

NEA gained independent charitable status and was so well thought of in "official circles" that the Energy Select Committee of the House of Commons was calling for more government support for its work, echoed in the same year by an independent report published by the Energy Efficiency Office itself. By 1988 over 500,000 homes had been draught-proofed by NEA Member Projects. NEA continued to consolidate its position as the professional voice of those fighting to alienate fuel poverty and, in 1990, in the face of open private competition, was awarded the Government contract to establish an organisation to administer the new £26 million Home Energy Efficiency Scheme (HEES). By 1994 the new organisation itself, EAGA (The Energy Action Grants Agency) (see below) was administering over £83 million of public money to alienate fuel poverty, rising to over £100 million by 1995.

7.2.2 In its earlier days, NEA had created and fostered local projects to carry out draught-proofing and insulation using the Community Enterprise Programme. In 1981 there were 28 such projects, by 1985 there were 100. With the demise of the Community Enterprise Programme, that provided its workforce, and with changes to Government funding for insulation and draught-proofing, that provided funding for materials, NEA had to fight its way through difficult years to maintain benefits for the fuel poor. Employment Training had very damaging results. Its emphasis on training, the loss of social welfare remit and difficulties in recruiting trainees resulted in the loss of many NEA Local Projects. After much lobbying - and after a tragic loss of impetus and continuity - the Secretary of State - then John Wakeham - finally introduced a new scheme of energy efficiency grants for low-income households that would be independent of Employment Training. It had been a damaging time for the fuel poor but NEA had not only won their battle but had established itself as the major consultant for Government in this sensitive area.

7.2.3 NEA provides training, advisory, developmental and representational services for its members. It pioneered and set-up City and Guilds qualifications in draught proofing skills in 1986 and has since produced many more training courses and certificated qualifications, including the Energy Awareness Certificate, used by many local authorities, housing associations and fuel utilities. It even became a requirement of British Gas' licence from its Regulator, Ofgas, that staff in its showrooms should have the NEA-developed certificate.

Since 1983 NEA has organised an annual conference. Its authority has been demonstrated by the regular attendance of Ministers of State. NEA also has its own quarterly magazine, "Energy Action" (See Section 5) and sponsors many authoritative seminars and discussion papers. A highlight of the NEA year in recent times has been "CONSERVENERGY WEEK" held at the beginning of the winter. It highlights the benefits of NEA's work to politicians, businessmen and local communities by involving them in a week of high-profile activity; usually featuring participants draught-proofing a property. In 1993 the Prince of Wales participated, joining an impressive list of Government Ministers and other community leaders who had assisted with "Conservenergy". NEA's work is best summed-up by its own description of itself:

"NEA is the national charity which develops partnership between central and local government, the private and voluntary sectors to tackle the heating and insulation problems of elderly and disabled people, single parents and other low-income households through a combination of practical action and policy development".

Contact address: Neighbourhood Energy Action,
St. Andrew's House,
90-92, Pilgrim Street,
Newcastle-upon-Tyne, NE16SG.
Telephone: 0191-261 5677.

7.3 ENERGY ACTION SCOTLAND

7.3.1 Although NEA operates throughout the United Kingdom and has many members in Scotland, there is also Energy Action Scotland (EAS). It was set up in 1983 and identifies itself as "a sister organisation to NEA" and is described thus: "Energy Action Scotland is the national charitable organisation promoting energy efficiency, energy conservation and affordable warmth for all". Like NEA, it attracts government funding, obtaining some 40% from the Department of the Environment and a lesser proportion from the Scottish Office. It has a wide range of business and local authority sponsorship and provides much the same service for its members and the fuel poor in Scotland as NEA does. Although the creation and continued existence of EAS may appear to be a strange and needless piece of nationalistic posturing, in fact things are different in Scotland - colder, for a start! - and EAS and NEA work well and creatively together. Each has the other's Director on its Board and there are Scottish fuel poverty projects that have membership of both bodies.

EAS has an annual conference and produces its own quarterly newsletter, "Energy Review". Contact address: Energy Action Scotland, 21, West Nile Street, Glasgow, G1 2PJ. Tel; 0141-226-3064.

7.3.2 Both NEA and EAS would describe themselves as "campaigning organisations" on behalf of the Fuel Poor but they are not alone in this endeavour. There are other bodies devoted solely to action for the Fuel Poor and others that, although having a wider remit, have some of their client group in the Fuel Poor category; for example, Age Concern, the Child Poverty Action Group, Care and Repair and many more.

Every year, over the winter, many of these groups, including NEA and EAS, come together with the Department of the Environment and the Department of Health to form WACH (Winter Action Against Cold Homes), which produces advisory material, including audio-tapes, and organises a "Helpline".

7.4 OTHER FUEL POVERTY PRESSURE GROUPS

7.4.1 More up-front in their action and, of course, independent of Government are such bodies as the National Right to Fuel Campaign, Right to Warmth (Scotland based), the Campaign for Cold Weather Payments, the Campaign Against VAT on Fuel, the Campaign for Cold Weather Credits and so forth. The National Right to Fuel Campaign is well known for its annual, "Warm Awards"; a light-hearted initiative but with quite a sting in its tail for those who are awarded, "The Blown Fuse Award" and "The Bad Smell Award", respectively for fuel poverty misdemeanours in the supply of electricity and gas! It is not an entirely negative event because there are also "The Warm Glow Awards" and "The Bright Spark Award", to recognise particular good services in the alleviation of fuel poverty.

7.4.2 The National Right to Fuel Campaign and the Right to Warmth work closely together, under much the same "intra-national" relationships as NEA and EAS, with joint conferences and joint publications. At present both are precariously funded by the private sector but achieved something of a land-mark in 1993 when they were each able to appoint a full-time salaried organiser. Their contact addresses are:

National Right to Fuel Campaign, 73, Collier street, London N1 9BE
Telephone: 0171-713-0300.

Right to Warmth, 21, West Nile Street, Glasgow. G1 2PJ.
Telephone: 0141-226-3064.

7.5 THE ENERGY ACTION GRANTS AGENCY (EAGA) AND THE HOME ENERGY EFFICIENCY SCHEME (HEES)

7.5.1 During the late Eighties the great movement to alleviate fuel poverty ran into difficulties, as Government policies played havoc with the established arrangements for financing the work and materials involved in draught-proofing and insulating the homes of low-income families. It is a complex and sorry story of dogma-led change whose casualties were the fuel poor and the unemployed. Under great pressure from the fuel poverty lobbyists, the Government included in its 1990 Social Security Bill provision for a new system of grants for low-income households. This was the Home Energy Efficiency Scheme (HEES), which provided grants for draught-proofing, loft-insulation and rudimentary energy advice. It had two dogma-led major faults - the method of contracting and a client contribution - but, in time, even these were removed, in the face of practicality and reality. At its inception the Scheme had an annual budget of some £26 million which, by 1995, had been increased to over £100 million.

7.5.2 To administer the Scheme, the Government invited proposals and tenders. The contract was awarded to NEA, which set-up an independent organisation, which became known as the Energy Action Grants Agency. Like NEA, it is based in Newcastle-upon-Tyne. To its great credit and, one understands, to the amazement of "official circles", NEA and EAGA were able to get the new organisation operational in only three months from the award of the contract. HEES became available on the 1st January 1991.

7.5.3 The Home Energy Efficiency Scheme was originally directed only at low-income households but in, 1994, a public out-cry over the introduction of VAT on fuel, following two humiliating by-election defeats the previous year, largely on that issue, forced the Government to extend HEES provision to those over 60. This, in turn, brought its own problems for the cash-limited Scheme. Even with an increased budget, the additional constituency - often not disadvantaged and certainly more articulate - made heavier demands on the available funds. ("Cash-limiting" means that provision is funding-led rather than demand-led). HEES was available to any householder who lives in a home for which the grant is available and who receives, or lives with a partner who receives, any one of the following benefit payments:

Income Support

Housing Benefit

Family Credit

Council Tax Benefit

Disability Working Allowance

7.5.4 In 1994, when the scheme was extended to any householder over 60 years of age, regardless of financial circumstances, strictly speaking, it ceased to be an exclusive "Fuel Poverty" measure. The scope of the Scheme was broadened again in 1995 when EAGA opened a branch in Northern Ireland, having won the government contract to administer the Domestic Energy Efficiency Scheme (DEES) - the Province's version of HEES.

Through HEES, money is provided to cover the cost of loft insulation, draught proofing and energy advice to a maximum cost, which is reviewed from time to time. Originally there was a compulsory client contribution but this was removed in 1994. It had been a matter of some note that HEES up-take was greatest where local authorities had arranged "Hardship Funds" to enable HEES providers to circumvent the client contribution requirement. Needless to say, these were not generally authorities of the same political persuasion as Government. Also in the original arrangements, besides a negligible D.I.Y. element, there had been two classes of provider: Network Installers and Listed Contractors. This had been a clumsy, bureaucratic attempt to meet the Government's wish for an element of client choice, which was finally abandoned in 1994. As things now stand, there is only one class of provider, Network Installer.

The Network Installer has an allocation of funds, can asses the client's eligibility for a grant and carry out the work. The Installer has to comply with a Code of Conduct designed to protect the vulnerable client-group and to ensure very high standards of behaviour and workmanship. After something of a struggle with Government, EAGA won the right to remove unsatisfactory installers, continuing its policy of providing a quality service for the client at a low administrative cost to the public-purse.

7.5.5 Although set-up by NEA, it is important to realise that EAGA is itself an independent, non-profit-distributing body. It has its own Board, though the Director of NEA is, at present, also the Chairman of that board. It works closely with NEA on matters of training and technical development and most of its Network Installers are also members of NEA. These members meet on Regional Forums that are administered by NEA; so there is very close co-operation at all levels.

Being a non-profit-distributing organisation; EAGA has created some operating surpluses as a result of its management skills and cost control.

In order to ensure that such funds continue to support the fight against fuel poverty, the EAGA Board established a Trust in 1993 to support research and other projects to clarify the nature, extent and consequences of fuel poverty. This is the EAGA Charitable Trust, which made its first grants during 1994. The trustees are mainly drawn from academics in this field and experts in fuel poverty alleviation. Grants are unlikely to be less than £1,000, with £5,000 being what the trustees described as, "a more realistic minimum" with about £25,000 being usual. The Trust's Administrator is, Bryan D. Emmet, Hayside Farm, Low Street, Sancton, YORK. YO4 3QY. Telephone: 01430 827552.

7.6 THE WIDER CONSTITUENCY OF FUEL POVERTY

7.6.1 The sort of organisations that benefit from funding by the EAGA Charitable Trust represent a remarkably wide-spread and extensive constituency of bodies operating in the fuel poverty area - remarkable, because this is not an area where there is a great deal of money available. For example, although there are purely commercial firms providing HEES, the majority of the providers are "not-for-profit" ventures, often part of a wider charitable activity. A dominant example of this would be Heatwise Glasgow, itself part of the wider WISE Group, and not only providing HEES but also acting as a training and development body in the fuel poverty area. At a further extreme, there are small organisations such as the Energy Conservation and Solar Centre (ECSC), with a staff of only two and a part-timer, chiefly giving a fuel poverty advisory service in the London boroughs but sufficiently expert to have won a contract from the Energy Savings Trust to run a LEAC (Local Energy Advisory centre). The Bristol Energy Centre (BEC) is a similar "not-for-profit" body, but operating on a much larger scale, with local authority backing and contracts for national level development work. It, too, won a LEAC contract and was also employed, by the National Energy Foundation, to produce the computer software that was to be used by all of the LEACs. A firm, such as Optima Services, Ltd., though much smaller than BEC and similar ventures, is run as a profit-making business as a result of a management buy-out from ECSC. It is therefore a heterogeneous area, with private mixed with "not-for-profit", a wide mixture of central government, local government, trust and private funding, size giving little indication of competence, scope nor professional status and much overlapping of staff, contracting and board membership. Among the "not-for-profit" organisations very few receive direct government funding, though many are helped by local authorities and carry out care-providing contractual tasks for them.

7.6.2 As far as central government is concerned, fuel poverty is chiefly the province of the Department of the Environment, mainly through the Energy Efficiency Office; though there is also interaction with Health and Social Security - for example, in contributory funding to the Winter Action Against Cold Homes initiative. In the present political climate it is probably a truism to suggest that those operating the alleviation of fuel poverty are needed more by Government than they need Government. The massive Government expenditure on HEES is more of a social security measure than an energy savings investment, though Government tends to list it among its environmental credentials. Without that expenditure and modest funding to bodies such as NEA and EAS, it would be difficult to see much official concern in this area. The excess of U.K. winter, fuel-poverty-related deaths over normal expected death-rates stands out starkly compared to rates in other European countries and there remain over 6-million households within the fuel poverty trap. It is therefore a major issue in the politics of energy conservation, even if it is not high on the political agenda. It is certainly to the credit of those operating in the area that the issue remains live and that even an avowed non-interventionist administration spends millions of pounds on it.

THE ENERGY MANAGEMENT MOVEMENT

8.1 For about a decade the Energy Management Groups were a jewel in the crown of U.K. energy conservation activities. Their history is certainly an object lesson in needlessly wasted opportunity and their demise was generally held, at the time, to be a particularly distasteful example of political nastiness. However, recent disclosures by retired civil servants suggest that it was more like a farcical, energy conservation re-write of "Murder in the Cathedral" than a deliberate piece of vandalism by a vindictive Under-Secretary of State. Whatever the real story, the U.K. energy conservation programme has been the loser and there are many valuable lessons to be noted by students of the politics of energy conservation.

8.2 In 1978 the U.K. Department of Energy launched the Energy Management Clubs but it was not long before the term "Groups" was adopted, rather than "Clubs", and it is under that name that the movement has since existed. The idea was to encourage like-minded people, with energy management responsibilities or interests, to meet together on a regular basis to exchange ideas - in much the same way as many other business and professional groups do; such as, accountants, transport managers, personnel managers and so forth. The EEO gave modest but adequate financial support to each Group for administrative expenses and provided further help through the Regional Energy Efficiency Officers (REEOs).

8.3 The movement grew rapidly; so that, by the mid-Eighties, there were over 70 Groups, representing all of the EEO's ten regions and with an individual membership of over 5,000. The greatest strength of the movement was that, from the out-set, it was given official formal access to the Under-Secretary of State who was responsible for energy conservation. This was achieved through the mechanism of a new body, the National Energy Management Advisory Committee (NEMAC), to which each Group was entitled to nominate a member - generally the Group's Chairman. This strength, now unfortunately lost, did much to enhance the standing of energy managers in companies and organisations. It also facilitated such innovations as the creation of an official energy management newspaper ("Energy Management"), the inception of an official energy management conference and exhibition (NEMEX) and the beginnings of official, formal training for energy management ("The Energy Managers Workshops"). Above all, energy managers had an official route to enable their experience, expertise and views to be made available directly to ministers and officials - even to ministers with officials by-passed! - and they were seen by chairmen and chief-executives of their companies to have this standing. It is a sad reflection on the present official attitude to this powerful resource opportunity that the latest EEO leaflet on the Energy Management Groups can go no further than to say that the groups are, "...recognised by the Energy Efficiency Office".

8.4 Before studying the events that led to what might now be termed "the down-sizing" of the Movement (For some in Government that meant, "Cutting NEMAC down to size"!), it is useful to record the activities of the early days. When the Groups were formed in 1978, the Under-Secretary for State for Energy, John Cunningham, made it clear that he wished the energy supply industries to remain at arm's-length, whilst providing support and encouragement. In particular, he had anticipated that those industries would be in a strong organisational position locally and would be constructively disposed to a government initiative (These were pre-privatisation days). The down-side of this situation was that they could well take-over the local running of the Groups (Chairman, Secretaries, premises etc.) and exert commercial influence. As energy managers have such a strong voice on fuel-choice decisions - if not actually making such decisions directly - the emerging Energy Managers Groups would have obvious commercial attractions to energy suppliers. The direct funding for administration provided by the EEO for the Groups was a low-cost way to keep predatory sponsors at bay. When that minuscule funding was removed in 1988, many groups were immediately snapped up by energy suppliers and became financially dependent on them. However, when the groups were first established and flourishing, there was a good relationship between them and the fuel suppliers, in spite of the mutually recognised dangers of commercial exploitation.

8.5 Because the Groups had such a strong and overt Government backing, they were able to thrive without direct financial support from local energy suppliers. They channelled into NEMAC able and enthusiastic members, who gave ministers and officials practical, even if often uncomfortable, advice about what was needed to achieve energy efficiency in U.K. industry, commerce and public administration. Groups drew their members from a wide range of business but, inevitably, a certain amount of special-interest grouping occurred; for example, a London Local Authorities Energy Management group was formed. To some extent this was counter-productive because cross-fertilisation was one of the original objectives of the Movement. Local authorities and other public bodies often had the most professional and well-trained energy managers in their ranks. Their experiences and skills could be useful; especially to managers in smaller businesses, for whom energy management might be only a part-time responsibility. Many Groups were heavily dependent - and still are - on their local government members for management and administration.

8.6 Very soon after its inception the Energy Management Movement began to articulate the need for formal training for energy managers. It was recognised that those with an engineering back-ground needed development in general management, financial presentation, financial control, staff motivation and so forth. Others, already from general management backgrounds, needed at least an appreciation of the technicalities of the hardware of energy management. In 1978 the Energy Conservation Unit (The forerunner of the EEO) of the Department of Energy launched its "Energy Management Courses", using the facilities of the School of Fuel Management, which was run by one of the publicly-owned energy suppliers. The Department of Energy gradually took over the organisation as a direct activity but, in 1980, it arranged to run the courses jointly with the British Institute of Management. They were well supported and became known as the, "D.En/BIM Energy Managers Workshops". There had always been a management board, drawn from a wide range of industry experts and enthusiasts, but this and the courses themselves gradually withered away towards the end of the Eighties, as official support and interest was withdrawn.

8.7 At about the time of the formation of the Energy Managers Groups the Department of Energy had introduced its annual conference and exhibition to promote energy management, the National Energy Management Exhibition (NEMEX). The conference rapidly became the national forum for the Energy Management Groups and the exhibition provided a prestigious shop-window for the newly-emerging energy conservation industry. Like so many other excellent initiatives, this, too, fell foul of the privatisation mania. First, Government encouraged a private conference and exhibition to rival NEMEX itself then, as the counter-productive folly of that became apparent, it let the official event go to the Energy Systems Trade Association (ESTA), who now run NEMEX as a private, commercial venture. The rival venture was left to wither. ESTA is an entirely competent and enthusiastic body of energy management enthusiasts from, as the name implies, industries whose concern is energy management. However, by privatising NEMEX government abandoned a valuable tool in its struggle to get U.K. industry, commerce and public administration to take note of energy management. Whatever else it did, it was an official signal to chairmen, chief-executives and to the energy managers themselves that government was serious about energy conservation. Besides the abandonment of this essential signalling process, the result of the "privatisation" was that the energy-saving purpose of the exercise was allowed to slip. Energy and fuel suppliers, energy purchasing consultants and others on the "supply-side" of the energy equation gradually took over. Strangely enough, though perhaps not surprisingly, the other major privatisations - of publicly-own energy industries - gave a new lease of life to NEMEX in the Mid-Eighties whilst the general run-down of U.K. industry put a strain on participation and attendance , in common with most other industrial conferences.

8.8 The urge to sell more in the artificially constructed competitive energy market brought the private energy "pushers" eagerly into NEMEX. At the same time one could detect a withdrawal of EEO interest and could witness a lessening of ministerial or senior official involvement. "NEMEX'94" finally signalled the end of any official real commitment to energy efficiency - let alone energy conservation - in the U.K.. Day One, traditionally the "grown-ups" day, was devoted entirely to energy supply, with the exception of the last session, which was concerned with water supply; not even saving water. The day was chaired by two symbols or icons of energy efficiency (The Director General of the EEO and the Chairman of ACE) but the programme spoke for itself:

Sustainable Development - Fuelling the Future.

Competition and Prices in Electricity Supply.

Competition in the Electricity Market.

Future Operation of the Gas Supply Market.

The Development of the Competitive Gas Market.

The Unfolding Energy Market.

The Challenge of a Consumer-Driven Market.

Developments in Water Services.

Day Two, the "little peoples'" day, was a real energy efficiency/ conservation event, with 22 presentations, in 3 parallel tracks, to tell the delegates how to save energy. However, being juxtaposed to Day One, its position in the pecking-order clearly relegated the matter of actually reducing energy-use to the Second-Division of business and official interest.

8.9 (Throughout its life NEMEX has always had a problem with its structure, being traditionally divided into a Top People's Day and a day for the others - by definition, lesser mortals than the Top People! The first day was, and is, the day for the ministers, chairmen, chief-executives and senior managers, being concerned with policy and top-management issues. The second day has multi-stream technical seminars. The conference dinner and the exhibition link the two and opening the exhibition to non-conference participants has overcome the problem of the technical sessions emptying the exhibition. However, the "class distinction" of the structure of the event remains unfortunate. It does not do much for the morale of the Energy Management Movement, that suffered such a blow when NEMAC was abolished in 1988).

8.10 Although the EEO continues to publish leaflets to promote Energy Management Groups, the Movement was effectively crippled in April 1988 when it was announced that the trivial funding for individual Groups' administration was to be withdrawn (A few hundred pounds per Group!) and the National Energy Managers Advisory Committee was to be disbanded. The whole sorry business was well covered in the press at the time, including the shabby way the enthusiasts were encouraged by EEO officials to try to maintain the Committee by private means whilst making it quite clear in official circles that it would no longer have official support, official status nor its privileged access to ministers. The final insult was EEO advice to the Energy Managers Groups to seek alternative funding from the fuel supply industries - the very supply-side sponsors that had originally been asked to keep at arm's-length from the Movement. Although some regional Electricity Companies stepped into the gap, the official British Gas instruction to its Regions was forthright, "British Gas considers that it would be counter-productive and give entirely wrong signals to energy management in the U.K. if the Company were to take-over the funding and support of Energy Managers Groups that has hitherto been provided by EEO and its Regional Energy Efficiency Officers". Whether this was righteous indignation or merely a cost-avoiding gesture can only be judged by the Company's subsequent commitment to energy conservation.

8.11 At the start, the Movement had risen rapidly to having 70 Energy Management groups. This rose to 80 in 1987. The individual membership had risen from 5,000 to 10,000. It is interesting that the latest official wording is that, "Over 75 EMGs have been formed throughout the U.K..." It is difficult to get an idea of how many remain active or how well their meetings are attended.

In 1983, Energy Paper 52, "The Fifth Report to the Secretary for State for Energy of the Advisory Council on Energy Conservation (ACEC)", endorsed the value of the Movement: "The Energy Managers' Group movement is considered to be one of the most cost-effective means of facilitating the transmission of information and provision of practical help to energy managers." (Para 156 e) This view of the minister's own expert advisers was insufficient to preserve the Movement's funding and official support. ACEC itself was destroyed shortly after that date.

8.12 To some extent, the Government's MACC ("Making a Corporate Commitment") initiative, launched in 1990, was a gesture to the remnants of the Energy Management Movement but it was aimed at the very top of company management. As such, it only obliquely helped the standing of the energy manager foot-soldiers. It was viewed as yet another Top Person's club, where ministers and senior officials could commune with "chaps like us". The energy managers themselves had created a more rigorous club (in spite of, rather than with, the EEO) in the form of the Energy Accreditation Scheme (See, Chapter 11). In addition, a group of enthusiasts had launched the National Energy Efficiency Association (NEEA) from the ashes of NEMAC, in an attempt to create a professional association for energy managers and to encourage a rigorous approach to training and qualification. A trivial amount of funding and active support from EEO would have harnessed this rich source of enthusiasm and expertise. Instead, to the obvious glee of some officials, it withered through lack of finance. The lack of official support for NEEA is extraordinary when one considers that the EEO had been working with the Management Charter Initiative (MCI), through the Best Practice Programme, to produce Standards for Managing Energy as a route to NVQ/SVQ awards. The NEEA, with its sights so firmly set on professional qualification, would have seemed an obvious, low-cost, valuable resource for government. On the other hand, without opening the unsavoury can-of-worms of the NVQ/SVQ concept itself, which could be seen as yet another device for avoiding the costs of rigorous training and examination that might result from EU Directives, it is not entirely surprising that officialdom preferred to ignore such a rival force as NEEA. After all, it was probably the perceived threat of NEMAC itself that led the same tribe of officials to engineer its demise.

8.13 In 1988, when NEMAC was disbanded, it was widely believed that the decision to get rid of it was a political one, emanating from the Secretary of State himself. Stories were circulating at the time that ministers resented the critical advice of the Committee's members and that the Under-Secretary had a poor opinion of the "no-hope, failed managers" that he saw as its members. It was believed that he had signalled to his officials that it should go, along with the Energy Managers Group funding that sustained it. ("Will no one rid me of this rascally priest?"!) More recent comment has revealed that the minister concerned had not intended the demise of NEMAC but the opportunity had been seized on by a senior official who thought he had, on his own admission, been brought in to close-down the entire Energy Efficiency Office, not only its Energy Management Movement. It is a sorry story and a lesson to those who seek official support that they need to cultivate and work with Whitehall officials for survival. Closeness to Ministers of the Crown is often not enough.

8.14 The potential of the Energy Management Movement, as exemplified by the concept of the Energy Management Groups, is substantial: especially as it exploits the penchant of the British to form voluntary professional and other interest groups. However, it is an uphill task to generate interest in energy conservation and to influence those for whom it is not an enthusiasm. It needs external resources to work. Most of all, it needs "official" backing. "Recognised by the EEO" is not enough. That very statement itself might be contrued as an insult to the hard-working enthusiasts. The potential of the Movement and the multiplier-effect value that it could produce from very modest funding was well demonstrated in the Early Eighties by the National Energy Management Competitions.

8.15 At about this time, even the then Secretary of State for Energy, Peter Walker, was warning that the U.K. had slipped from its position in the World as a leader in energy conservation activity. Declining interest in the Energy Management Groups was part of that slippage, with many of them struggling for survival. The Dept. of Energy officials who were responsible for the Groups were looking around for a way to re-invigorate the Movement. The electricity and gas supply industries were approached, to ask for their technical staffs to become more active in support, but they found it difficult to comply without turning Energy Management Group meetings into sales battle-grounds for the competing fuels - a danger which the original "arm's-length" policy had rightly recognised. In their pre-privatised form the industries were able to respond positively but more was needed, for a successful re-invigoration process, than merely increased numbers of technical lectures, demonstrations or generously-provisioned site visits. Although an imaginative and effective initiative was found, it should be noted that it was wholly-funded and mostly resourced by the sponsoring publicly-owned fuel supplier. The EEO's disquiet about the Groups did not actually run to producing real public money for a remedy. This failure ensured that any initiative that was devised could only have limited life and could only survive so long as commercial advantages continued to accrue to the sponsor.

The successful initiative was the NEMAC/ British Gas National Energy Management Competition. This was similar in concept and operation to a computer-based Management Game, a training device that had become well established by the beginning of the Eighties. The School of Fuel Management at Solihull had developed a computer-based simulation of the energy use of a medium-sized factory; enabling users to experiment with various energy-saving strategies and to calculate the cost-saving benefits. The sponsor paid for development work to introduce a competitive element so that users could become "players", competing to maximise energy savings and cash returns on energy efficiency investments. It was particularly suitable for team-working, with teams composed of experts with various management skills - for example, energy process engineering, accounting, statistical control and so forth: the very mix of people generally found within Energy Management Groups. NEMAC negotiated with the Dept. of Energy to adopt the competition, arrange ministerial involvement and on-going coverage in "Energy Management", the official government newspaper for energy efficiency. The greatest value to the Movement was that teams could only be entered by Energy Management Groups; thus producing an immediate boost to individual membership, re-inforcing Group loyalty and giving the Groups themselves a new and useful interest.

8.16 The first competition ran in 1983. Eighty-two teams entered, of whom sixty-two persisted for the whole five-month period of the competition. Even before the end NEMAC was asking for a repeat run, in response to pressure from the membership. An Under-Secretary for State for Energy presented the prizes (Personal Computers) at the 1984 NEMEX Conference and announced the start of another round, this time based on a computer-based simulation of a complex of commercial buildings. The Energy Management Movement had certainly benefited. At least two papers are known to have been given, by invitation, to European conferences and the sponsor was able to turn the software into marketable packages, including overseas sales. That the competitions were stopped after the second run, was a matter of fuel-industry privatisation and the EEO's attitude towards NEMAC. That aside, the episode amply demonstrated what could be done with a valuable resource of goodwill and personal commitment generated with officially encouraged and adequately funded Energy Management Groups.

8.17 According to the EEO there are still "over 75" Energy Management Groups "recognised". They may be contacted via the relevant Regional Energy Efficiency Officer, whose name and contact details appear in the latest issue of, "Energy Management". That or the NEEA, will put one in touch with the Energy Management Movement.

Apart from meetings of local Energy Management Groups, NEMEX and Making a Corporate Commitment events, there is a more general life after the death of NEMAC for energy managers. The EEO has been organising its "Roadshows" and other energy efficiency seminars around the country and there are the National Energy Awards organised annually by the NIFES Consulting Group and the Major Energy Users' Council - not to be confused with the National Energy Manager of the Year (ESTA), the recently defunct Gas Energy Management (GEM) Awards of British Gas nor the PEP and BETA Awards of the Electricity Industry. Many universities and colleges have energy management-related courses, ranging from short CPD events to full Masters Degree courses, and there is quite an active local trade in seminars on energy efficiency - often now presented in environmental protection terms - either run by local branches of professional bodies or as commercial ventures by training organisations. However, where it is well supported and administered, the local Energy Management Group offers the best entree to support, general interest and training for the energy manager, student or researcher in energy conservation in industry and commerce. The fact that an energy buyer can make savings of about 20%, even before energy efficiency is considered, may induce industry to re-appraise the status of "Energy Management".

9 ENERGY ADVICE, THE ENERGY AUDIT AND ENERGY LABELLING

9.1 Advice, audit and labelling are issues in energy conservation that are interconnected, sometimes pursued collectively or, at other times, individually. They are defined, variously, as parts of a single process or are capable of stand-alone discussion. They have all suffered from misleading definition or application. For some observers, they offer market opportunities as products with profit potential; for others, they are some sort of public service or just means to other ends.

They are interconnected but it is useful to examine them as separate phenomena to understand the many political, social, administrative, legal and technical issues that they involve for the energy conservation activist or policy-maker.

9.2 ENERGY ADVICE

9.2.1 "Energy Advice" has been a major element of energy conservation policy for most administrations. It most usually refers to an activity for householders; though, of course, it is also required by industry and commerce. However, it is to the domestic sector that politicians and energy conservation lobbyists generally refer when using the term. It has been a declared element of government policy for many years and has even appeared as positive, funded action in the Government's Home Energy Efficiency Scheme (HEES), where a fee of £10 (1995) is paid to a Network Installer to give the HEES recipient "energy advice". The Energy Saving Trust (EST) has also set up a network of Local Energy Advice Centres (LEACs) under a scheme funded by the Energy Efficiency Office and involving the National Energy Foundation.

9.2.2 There is a thriving industry involved in giving "Energy Advice", quite often funded by local authorities for their tenants; plus an equally thriving industry involved in training energy advisors. A great boost was given to the activity in 1992, when the Gas Industry Regulator, OFGAS, made it a condition of operating that British Gas should make energy advice freely available in its showrooms and that each showroom should have at least one adviser trained to the level of City and Guilds Certificate in Energy Awareness. Most Regional Electricity Companies (RECs) have followed a similar path, voluntarily.

9.2.3 All political shades of Government in recent decades, both in the U.K. and elsewhere, have recognised the need to give energy advice. On the whole, people will want to do the right and sensible thing but they need to have it explained to them whenever there is an element of technical understanding involved; especially if there is also technical controversy and risk. Cavity wall insulation is such a point of controversy, involving significant cost and the remote risk of damaging what is for most householders their most valuable asset. Similarly, at a trivial level, people wonder whether it is environmentally worth the bother of switching off a light in an unoccupied space. These are the sort of technical questions people might ask who are aware of energy conservation measures. Many more may not even be so aware and will need advice, simply, on what can be done; that is to say, that cavities can be filled or lights can be switched off - whether or not either is a Good or Bad Thing. Where grants exist, advice is needed to make people aware of them, how to apply, who is eligible and so forth. These needs seem to be generally agreed. What is less clear is how this advice should be delivered; the "Energy Advice Centre" is merely one solution. Disagreement arises over who should pay for providing the advice, in whatever form it is provided. Should it be delivered as a public service or as a commercial enterprise - or as some mixture of the two?

9.2.4 Given the level of official activity in this area, it is surprising that it has not been better researched and developed; for example, with the level of fundamental research that has assisted energy conservation teaching in schools. Such research as has been published suggests that there is much to be done. The City and Guilds Certificate, that OFGAS imposed as a standard on British gas, is purely what it says, "Energy Awareness", with insufficient content concerned with the very difficult task of turning the knowledge into effective advice. The same certification is the only basis for installers to be able to deliver the £10 advice element of HEES. This is a step in the right direction but woefully inadequate. Fortunately, most of those delivering this service do more than the small fee actually covers. Most are also acutely aware of the shortfall.

9.2.5 On the whole, HEES and, even the OFGAS approach, has tended to emphasise the Fuel Poverty aspects of energy advice. Few services are available for what some commentators call the "fuel rich"; that is to say, householders who are quite able to meet their fuel bills, able to finance energy efficiency measures but who lack guidance about what to do. It is an area that is even less well researched than advice for the Fuel Poor. However, a series of studies by Projects in Partnership, initiated over

1994/95 by the Energy Saving Trust, has advanced understanding and could be the basis for more. The most recent innovation in energy advice - for all types of household - has been the Government's own Local Energy Advice Centre (LEAC) venture.

9.3 ENERGY ADVICE CENTRES

9.3.1 Before the present government-funded LEACs came into being, there were, and still are, many local or wider initiatives to provide points-of-contact at which householders may seek advice on how to improve their energy efficiency and reduce their fuel bills. Given the high cost of one-to-one advice to people in their homes, it was natural to seek more cost-effective ways of delivering advice, beyond the obvious method of leaflets and posters. The concept of centres where advice could be delivered to visiting clients seemed to provide a less costly compromise but it has yet to be shown that even the "Advice Centre" is a cost-effective form of advice delivery. Recent work by Neighbourhood Energy Action and others has suggested that the most effective form of energy advice delivery is still one-to-one in the client's own home. Though apparently expensive at the point-of-delivery, in the long run the effectiveness may prove that it is also the most cost-efficient. Further research is needed. Meanwhile, the "Advice Centre" enjoys some general credibility and has become a well-established public service offered by several forward-looking local authorities, either with or without the LEAC initiative.

9.3.2 HESACs AND THE BIRMINGHAM EXPERIMENT

It is not widely remembered - or, perhaps, it is conveniently forgotten - that, around 1978/79, the Energy Conservation Unit (ECU) of the Department of Energy was working on a plan to set up "HESACs" (Home Energy Saving Advice Centres). The first two had been planned for Liverpool and London, in association with the Building Design Centre Trust, and some staff had been recruited. In October 1979, John Moore (the then new Tory Under-Secretary of State for Energy) announced that the project had been dropped. It is one of the ironies of energy conservation history that it was to the same John Moore, by then Lord Moore and Chairman of the Energy Saving Trust, to whom Government entrusted management of its "new idea", the LEACs (Local Energy Advice Centres), set up by the Trust to deliver home energy saving advice.

9.3.3 As far as the politics of energy conservation are concerned, the history of the notion of delivering energy advice from local centres is yet another example of a rule: "It is not the value of the idea that matters but who can claim ownership." The HESACs were only one of many initiatives that were closed down in 1979/80 merely because they were the ideas of the "Other Side". Like some other ideas, HESACs have resurfaced but, in their case, with the additional problem of having to present a blatant piece on interventionism as a free market initiative - a problem that hinders so much progress in energy efficiency under the sort of political thinking that gained the upper-hand from 1979 onwards. A trick for the energy efficiency lobby is, therefore, to help whatever politicians are currently in power to present commonsence good ideas in terms of the current political dogma and to re-package old, well-tried programmes as new, radical initiatives.

9.3.4 Up until 1979 the U.K. cruised along with a general consensus that energy conservation was to do with public service and civic need and that it was entirely reasonable to fund such services from the public purse. From 1979 the will or inclination to govern and serve in such terms was not part of the make-up of those who took decisions about energy conservation. As a consequence, the provision of energy efficiency advice suffered along with many other public duties. The HESAC project was an early casualty but it was merely sharing the fate of the Energy Quick Advice Service (EQAS), the Energy Conservation Scheme (ECS), the Energy Survey Scheme (ESS), the Industrial Energy Thrift Scheme (IETS) and many other initiatives that were directed at advising energy users in the non-domestic sectors. And, to these earlier casualties, must be added the recently discontinued Energy Management Assistance Scheme (EMAS).

9.3.5 The Home Energy Saving Advice Centres (HESACs) were developed following an extensive and well funded experiment by the then British Gas Corporation, as a response to a direct request by the Under-Secretary of State for Energy. The fuel suppliers were a development resource then available to Government, arising from the special relationship that public ownership of natural monopolies made possible. Where energy conservation was concerned, the publicly-owned coal, electricity and gas industries had earned a reputation for a fine record of assisting and supporting government programmes; in marked contrast to the fragmented and limited reactions of both U.K. and international oil companies. The British Gas experimental energy advice centre in Solihull, wholly developed and funded by the British Gas Corporation, was a classic example of pre-privatisation fuel industry/government co-operative development. It provided experience and lessons in an area that was largely reliant on speculation. The behavioural, particularly marketing, aspects of energy conservation then, as now, were poorly researched, in contrast to the extensive R&D effort on hardware aspects of the matter.

9.3.6 A prominent shopping-centre retail unit was rented in the busy pedestrian thoroughfare around the Bull Ring in Birmingham, one of the leading retail centres in the U.K.. On the ground floor were explanatory displays of energy conservation measures, demonstrating what could be done to save energy in the home and how householders could take action. There were also display examples of government and other advisory organisations' literature, videos and films, plus examples of energy-saving devices and materials. There were advisers on hand, including one heating engineer provided from a rota of practical district heating representatives by the local Region of the British Gas Corporation. On the upper-floor there was a small demonstration theatre and lecture room.

9.3.7 The advice centre was maintained for about two years, on its original site, and then absorbed, because of financial restraints and changing market circumstances in the Early Eighties, into an upper-floor space in a major gas showroom in a nearby shopping centre. If nothing else, the move demonstrated the value of a street-level, prominent site to attract domestic energy users because client traffic declined as soon as the Centre was, literally, "kicked upstairs", banished from street-level retailing space that could be more profitably used.

9.3.8 The main lesson learned was that the provision of home energy advice by personal interaction at such a centre is a very expensive undertaking. A high-street or retail-centre location is essential but that, in itself, involves high overheads. The staff who are giving advice need to be well qualified and of high calibre. (Later work, in the mid-nineties, by Neighbourhood Energy Action among low-income households, has suggested that such person-to-person advice is even more effective when given in the person's home. It would be useful to have a research programme to assess the relative cost-effectiveness of the centre approach compared to the own-home approach.).

The pattern of enquiry traffic at the Birmingham centre largely followed what one would expect in retail traffic terms but Saturdays revealed an unexpected phenomenon; that of small-business people calling in for energy advice relating to their business premises and processes. It is interesting that this element, as well as the overall HESACs experience, appears to have been considered in the latest, LEAC, initiative

9.3.9 LEACs (Local Energy Advice Centres)

The trials, tribulations and works of the Energy Saving Trust are covered in Section 6. One of its earliest problems was that the Gas and Electricity Members were annoyed that, having been introduced into the venture on the understanding of a tripartite (Gas, Electricity and Government) arrangement and share of cost, there was little evidence of real Government cash. This was particularly irksome for British Gas and Ofgas, as the gas customers' contribution was quickly reaching figures involving millions of pounds. Fortunately for Government - or, at least, for its "face" - the Energy Efficiency Office had on the stocks a scheme to set-up local energy advice facilities under a shared funding arrangement involving both public and private money, roughly on a fifty-fifty basis. Instead of going ahead in the usual way (either running the scheme from the EEO or contracting directly to the service deliverers), Government launched the venture as an Energy Saving Trust scheme; that is to say, the EST was given the fund, which it could 'top-slice' as a source of income, and was then left to seek a contractor to manage it. In the event, it awarded the contract to the National Energy Foundation (NEF), which, in turn, sought local deliverers on contract and contracted-out the task of developing a common computer-based data base. However annoying this might have been to the EEO officials, surveying a long and costly sub-contractual trail, it got the Government "off the hook" concerning its visible cash commitment to the EST and a fairly well funded, large-scale experiment in the concept of the "Energy Advice Centre" came into being.

9.3.10 As a research and development project the LEAC venture is long overdue and, because of the wide diversity of participants and the involvement of very many experienced energy advice deliverers, it may well prove to be a very valuable investment. In its 1993 Report on Energy Efficiency in Buildings, the House of Commons Committee on the Environment had commended the concept but chiefly as a channel for funding by the utilities. ("We believe that local energy advice centres provide energy utilities with excellent opportunities for promoting their energy efficiency initiatives among their local customers. We recommend.....the RECs and British Gas regions to contribute to the running of them").

In the long run, the main value of the LEACs may be the information they provide to policy-makers in the future, rather than their success in delivering energy advice or in delivering actual energy savings. In the first place, privatisation dogma has sown the seeds of failure; in spite of a painful lesson that Government should have learned from its attempt to act similarly with CREATE (The Centre for Research Education and Training in Energy) - (See Section 12. The Education Sector.). Only short term "seed corn" funding has been provided for the LEACs. After two years they are to "stand on their own two feet", raising funding from client and/or industry and commerce - sources that are notoriously uninterested in energy conservation and whose lack of willingness to pay for advice was why a subsidised service had to be started in the first place! Even the seed-corn period means that the Centres are competing in an unfair, subsidised manner with existing energy advice ventures; creating all the ill-will that might be expected.

9.3.11 A further problem is that, in the search for sponsorship for the Centres, Government encouraged the Centre managers to seek it from fuel suppliers and heating service industrialists. Apart from bidding to take on a centre, those bidding had to match Government funding, pound-for-pound, with locally raised funds. As the Centres are supposed to offer independent advice, what of independence if such local funding comes from individual energy suppliers or were to come from suppliers of one specific make of energy-saving or energy-using equipment? Furthermore, in practice, a large proportion of such local funding that has been raised has come from local authorities, which makes a nonsense of any claim that the Government's bizarre approach is removing from the public-purse the cost of delivering energy saving advice. Even so, the pound-for-pound was difficult to raise because, in recommending direct funding by "the utilities", the Commons Committee had not grasped that gas and electricity in the U.K. were no longer utilities but private, profit-seeking companies; nor had they grasped that the Regulators - OFFER and OFGAS - had other views of such uses for customer and shareholder cash.

9.3.12 Finally, problems will arise from the fact that some of the bodies that have been successful in bidding for the Centres are inappropriate or too narrowly focused. The involvement of such bodies may be useful in providing research information but leaves questions over the quality of or appropriateness of the advice and raises doubts over commitment across the whole range of target markets. At one extreme, the wholly-owned subsidiary of a Regional Electricity Company has won a Centre. At another extreme, a charitable body, concerned with the Fuel Poor, has a Centre. Where, one might ask, in this situation is there any confidence in the independence of advice for the ordinary householder or in the expertise of advice for the local small business?

9.3.13 The long contractual chain of the LEAC venture has already been mentioned. Apart from the sheer waste of effort and dissipation of public money, the extended chain of command cannot help the operation: EEO, Energy Saving Trust, National Energy Foundation and Local Operators, some of which have, themselves, added a further link by setting up a subsidiary operation. At the time of writing, two LEACs have closed down and others are facing matching-funding problems for their second year of operation. "Cost-per-contact" and "Cost-per-kilowatt-hour-saved" will be an interesting measure of this venture. At least the extended chain of contract and command has created many new administration jobs. Whether Government regards these as "real jobs" has not been said.

9.3.14 The main commercial problem with delivering domestic/ household energy advice is that it is expensive, even to do badly, and therefore costly to purchase. Householders are not only disinclined to pay for most forms of professional service but are even less motivated to pay for a service that is costly and seems only likely to tell them that they have problems. Most service industries (fitted kitchens, fitted bathrooms; even double-glazing) have to include surveying and advice in the overall product price rather than charging for it as a discrete, stand-alone service.

Specific energy advice, to individual householders, is unlikely to survive as a product in its own right but probably needs to exist as below-the-line marketing support for more tangible products or as a by-product of energy rating or labelling.

9.3.15 An opportunity that seems to have been missed in energy advice development in the U.K. is the "Feedback Energy Meter". Work in the U.S.A. in the late Seventies, brought into the U.K. by some L.S.E. researchers some years later, indicated that savings in the order of 15%-20% could be achieved by behavioural changes alone when feed-back energy meters were fitted in communities. To a certain extent, the recently arrived ECOFEEDBACK movement is tapping a similar motivation.

The idea is that all energy use - for example: oil, gas and electricity - is electrically fed into an intelligent meter that electronically produces what is termed "log-pile" and "speedo" information to the house-hold, constantly, in real-time and expressed in money values. The meter is prominently positioned at eye-level; for example, in lobby, kitchen or even living-room situations.

The "log-pile" information means showing cumulative consumption against budget, against comparable times,corrected for temperature and so forth. The consumer reacts as one might in more primitive times, by seeing the "log-pile" of energy purchased, or planned to be purchased, diminishing with usage. The modern problems of long time-lapsed billing, annual-spread billing and even out-of-sight metering, which shield users from the immediate impact of energy cost, are removed.

The "speedo" information means showing just how much an appliance or comfort-level is costing - per minute, per hour, per day, - in money terms.

The consumer can see the immediate effect of, say, closing a window, turning-down a thermostat, turning-down a fire, half-filling a kettle and so forth. Rival fuel suppliers have a strong interest in this because comparative running costs become very obvious to the users.

Energy utilities and their closely-associated meter manufacturers have always claimed that such devices are prohibitively expensive, technically difficult or just not effective. A company brought a device from the USA in the late Seventies that, at that time, would have cost about £15 and actually avoided mains-electricity connections. The story of the Dept. of Energy's failure to take the utilities along with the idea, of industry obstruction and the eventual demise of the initiative would be a depressing revelation now but technology has moved on since then and there are supposed to be competitive forces at work within energy supply. It is interesting to speculate on what a predatory supplier could do to a competitor by supplying such devices into a rival supply area. Early evidence showed that massive savings could result. A determined government could reduce energy consumption dramatically and, in the resulting demand collapse, re-acquire the privatised fuel-suppliers at knock-down prices, perhaps giving them to the local and democratic control of local authorities!

9.4 THE ENERGY AUDIT

9.4.1 The "Home Energy Audit" in the U.K. has had a long and chequered history. The term itself is used, loosely, to cover a range of activities and services: as a procedure for delivering energy advice, as a scientific measuring tool, as a means of assessing what measures could be taken to improve a building's energy efficiency or as a tool for producing an energy label or an "energy rating". Although it is now generally mixed up with energy labelling, the "energy audit" evolved as a step beyond general advice and, for a time, various attempts were made to market it as an advice product; either directly to "the fuel rich" or via landlords for the "fuel poor". Its mechanics have ranged from a simple manual check-list, against which a dwelling is checked, to computer-based calculation models of varying degrees of sophistication. In essence, an energy audit is merely the process of finding out how much energy a household uses and whether that energy is used efficiently - not to be confused with "effectively". (The uninformed householder will often confuse the two terms and assume, for example, a gas-fire salesperson is explaining how much heat an appliance gives out, when speaking of its "efficiency", rather than explaining how well the appliance converts purchased gas into delivered heat.)

9.4.2 In the Seventies, at the request of the government at the time, British Gas put together a service for its customers which, today, would be recognised as an "energy audit". The idea was that a qualified heating engineer would call at the house and carry out a survey of the structure and of the heating and hot-water systems and appliances to make suggestions where opportunities for energy-saving existed. This would be an entirely free service under the strap-line, "Don't Waste Your Energy!". A range of energy-saving products and replacement, more-efficient appliances were on offer - from roof-insulation to modern gas-boilers. This was, in fact, the first "one-stop" energy conservation service ever offered to the public in the U.K.. The free energy audit was extensively advertised in two of the Company's operating Regions. In one, a "DIY Mart" approach was added with a cash-and-carry facility at town-edge depots on Saturdays, plus showroom-based "Advice Centres". Commercially, it was a total failure. Take-up of the free energy audit did not get into double-figures in either trial area. Product sales were abysmal.

9.4.3 Shortly after this attempt, the Open University, in association with the newly-launched TV Channel Four's, "For What It's Worth", offered a simple DIY check-list energy audit to the public, which was one of the earliest computer-based schemes. For the modest fee of £6, the O.U. sent the householder a form that led through a series of questions about the dwelling's structure, heating systems, occupant behaviour and past fuel consumption. This was returned to the O.U. for processing, which produced a print-out of energy efficiency and proposals for energy improvements. The take-up was reported at about 10,000 - in spite of the national T.V. exposure. No follow-up research has been available to assess whether viewers who paid the fee actually took any subsequent energy-saving measures. The scheme gradually faded away.

9.4.4 All the time, from the Seventies to the Early Eighties, computer-based energy audit programmes were being developed; some for use on early lap-top computers, others for processing by the initiator on an office-located machine. In the Mid-Eighties the Energy Efficiency Office again asked British Gas to conduct a pilot marketing exercise to support a new company named HEAAT, (Homes Energy Audit, Advice and Treatment) Ltd., which was to offer a Home Energy Audit on a fee-paying basis, plus follow-up remedial services and a certificate of energy efficiency. British Gas advised the Dept. of Energy strongly against the venture but political pressure prevailed. A test-market area was defined, with a target market of 25,000 households. In spite of extensive marketing on T.V., radio, local press and Gas Showrooms, only about half-a-dozen householders responded. To make matters worse, the computer programme failed to deliver print-outs within the specified period and no follow-up sales were recorded. HEAAT, Ltd., faded away and the cost to British Gas was considerable. Certainly, no energy was saved!

9.4.5 A few years after this experiment a new development occured that promised to enable the Home Energy Audit to work. This was the arrival of the concept of Energy Labelling or Energy Rating. The energy audit was now able to be seen, not as an end-product, but as a means to obtaining an asset that had commercial value, that could enhance the re-sale or lettable value of the dwelling. The HEAAT, Ltd. venture had trailed the idea, with its certificate, but not in a form that suggested absolute values. The first in the field was the National Energy Foundation (NEF), with its National Home Energy Rating Scheme (NHERS). A rival initiative (MVM "Starpoint") followed shortly, which indirectly led to the official government SAP (Standard Assessment Procedure) rating.

9.4.6 Meanwhile, the now-privatised Gas Industry was led by its Regulator (Ofgas) down the failure-strewn path of energy-audit-and-energy-advice that government had forced on it in earlier times. At about the same time a number of Regional Electricity Companies were renaming their district salespeople, "Energy Advisers" and offering their customers what was called an energy audit service, but to sell electric appliances rather than to offer energy-saving services. British Gas had also been forced by Ofgas to train showroom staff to give energy conservation advice but the Regulator still required a trial of a sort of "out-reach" service, beyond what was on offer to show-room callers. The Ofgas initiative, in 1991/92, required British Gas to trial an energy audit and energy-saving products marketing venture in two of its districts, a part of Scottish Region and a part of North Eastern region. The target this time was neither the "Fuel Poor" nor the "Fuel Rich", but those on the margin. Like previous ventures it was a total commercial failure. A reputable marketing consultancy was employed and both British Gas and a chosen business partner had very powerful political reasons for producing a success. There was even a targeted mail-drop and a participators' competition, involving free gas appliances. Few took-up the offer, scarcely any energy-saving products were sold and, again, no energy savings resulted. However, follow-up research did show that members of the public had become more aware of energy conservation! It is also admitted, by all parties concerned, that there were organisational and attitude failures; some connected with post-privatisation morale and disorganisation, others from a mismatch between the national company's staff and organisation and those of the smaller partners. The scheme was also deficient in that it did not offer an energy label or rating, which, by now, had become more accepted as a product in its own right. Perception had moved on since the "Don't Waste your Energy" and the HEAAT days, which the scheme had not recognised.

9.4.7 The MVM Starpoint and NHERS schemes had begun to produce follow-up energy-saving product sales through their Energy Labelling or Energy Rating approaches and the MVM Starpoint scheme had planned commercial success, to a large extent, based on such additional revenue. However, this remained at fairly modest levels, "leads" proved hard to sell to follow-up service suppliers and the "Energy Label" itself was being seen as the new product, which could "pull" the audit and remedial sales after it.

9.5 ENERGY LABELLING/ENERGY RATING

9.5.1 The idea of "energy labelling" dwellings - or any building, for that matter - has become established in the U.K. over the last few years and has become technically respectable; chiefly as a result of the Government's SAP (Standard Assessment Procedure). This is now part of Building Regulations and was forced into the public arena by confusion over the existence of two rival methods of assessment and energy rating. ("Energy Rating" and "Energy Labelling" are synonymous but usage seems to be applying "rating" more and more to buildings, leaving "labelling" mainly to appliances).

9.5.2 It was as long ago as the late Seventies that the Buildings Group of the Advisory Council on Energy Conservation (ACEC) was getting to grips with energy rating and formulating the notion of legislation, to require all buildings offered for sale or rent to declare an energy rating. This had been done successfully for motor vehicles and it was widely expected that it could be done for buildings. Computer-based models and manual methods for assessing a building's energy demand had been available since the Late Sixties; largely from basic algorithms developed at the University of Newcastle-upon-Tyne. Central-heating installers had had acceptable calculators for assessing heat-loss for decades; even though these tended to oversize the installations. The main reason why building energy rating did not develop nor become accepted by the professions, the trade and the public was that, where buildings are concerned, the energy activists had allowed enthusiasm for prescriptive approaches to be developed; demanding qualitative, judgmental schemes to be proposed, rather than the simply indicative scheme that has had such a profound effect on the energy efficiency of motor-cars.

9.5.3 In the Early Eighties it seemed that the government and the housing industry had come close to reaching agreement on the energy rating of new dwellings. Many major house-developers and building societies were prepared to sign-on to a scheme but the Secretary of State concerned suddenly proposed presenting it with "Star Ratings". A simple numerical scale, such as the present SAP produces, was too mundane. He needed something that would make its mark with the media. It is to the credit of the officials that they accepted the expert opinion of the experienced marketers and behaviourists, that a scheme that implied that certain levels of rating were better, absolutely, than others would disquiet the trade and industry. The politician was not to be so persuaded and it was no surprise that the "Star Rating" scheme quietly faded. House-developers quickly recognised that a "Two-Star" house would be harder to sell than a "Three-Star" and that the difference might merely reflect whether identical building were at the top or bottom or on the south or north face of a hill. Lack of an energy "Star" might destroy the added value of a view!

9.5.4 After the non-event of the "Star Rating" scheme and the various commercial failures of energy audit initiatives, the energy rating concept did not progress until the Milton Keynes Development Corporation developed its National Home Energy Rating Scheme (NHERS). The Scheme has a methodology to produce an assessment of a dwelling's energy efficiency on a scale 0-10, where '10' represents the most energy-efficient home. The Foundation licences surveyors to carry out energy surveys and to use a number of software programes that offer varying levels of home energy rating assessments, indications of the annual energy consumption, plus calculations of CO2 emissions. It also evaluates, and recommends, remedial measures. At present the service is making little impact on individual householders but is being purchased chiefly by housing developers, fuel suppliers and landlords (local authorities and housing associations).

9.5.5 At about the time that the NHERS was being developed, there was official encouragement for a rival scheme, being developed as a purely commercial venture. This was the MVM "Starpoint" Scheme. In free-market terms this was a Good Idea but it mainly created confusion in the minds of the public and brought out the worst in the competing fuel suppliers, vying with one another to cultivate whichever rating method showed their product in the best light. The NHER scheme had been "owned" by the Department of Environment, whilst "Starpoint" was a Department of Energy initiative. This may have had some bearing on the fact that such a ridiculous, unnecessary competitive situation emerged.

The Starpoint Scheme also has trained assessors and expresses energy efficiency performance in a "star" format: One star represents poor performance, rising to five stars for the best. Like the NHERS, it produces and evaluates recommendations on energy efficiency improvement measures; in fact, at its outset, commission income from such follow-up work was to be part of its commercial structure. The operating company was offering "leads" from its surveys to energy suppliers and appliance installers but there are no reports of much success in this direction. The assessment also added CO2 savings to its reports.

9.5.6 In the Early Nineties trade and professional pressure for some sort of reconciliation between the rival schemes led to EEO action and discussions with the two providers. There was little progress until the impasse was broken by the appearance of the Government's own Standard Assessment Procedure (SAP), developed by the Building Research Establishment (BRE) and based on its BREDEM (BRE Domestic Energy Model). It assesses a dwelling's energy efficiency on a scale of 1 (poor) to 100; as such, it is more flexible and credible than the NHERS 11-point scale or the Starpoint five. In a Perfect World, the SAP would make other rating schemes redundant and was long over-due. Some commentators wonder just how the politics operated which allowed the confused "market" situation to arise in the absence of an "official" method and scale, which appeared suddenly when the situation began to generate criticism of government.

The situation now is that private energy rating schemes, whether institutional, such as by NEF, or commercial, such as by MVM Starpoint, can be "authorised" to relate their ratings to the SAP.. At the time of writing, two new players have entered the market and are seeking authorisation; these are Elmhurst Energy Systems, Ltd. and HESTIA, Ltd.. The EEO is inviting others to seek SAP authorisation.

These developments indicate yet a further failure of officialdom to appreciate how markets work and the confusion and waste of resources caused by piece-meal tinkering through ill-digested understanding of competition. Had government taken a lead by requiring the energy-rating of dwellings and by producing an official standard assessment method - or even just producing the method and merely recommending individual take-up - the "market" would have created the means of delivery. Market differentiation would have arisen as a natural adjunct of the competitive imperative and would have taken the form of levels or approaches to delivery services. Instead, differentiation appeared in the form of variations in the product, so that the consumer was confused. The product failed to make an impact on the market and left potential users worried that the wrong version might be bought. This left the market as it now is, stagnant.

In 1988 a European Community Directive had required energy rating but it had suffered a U.K. veto, arranged by the then Under Secretary of State for Energy.

9.5.7 It is not too late. If legislation were to be introduced to require all dwellings offered for sale or to let to state a SAP rating, what should have been the original market will be created. All players in the assessment business will benefit, the public will recognise the product and the suppliers will be able to exercise their marketing and competitive talents by offering service excellence, by varying methods of delivery, "bargain" approaches and so forth. The more conservative and timid customer would still be able to opt for the "safe", institutional service offered, say, by NEF or other not-for-profit organisations and a real "free market" will exist, suiting all tastes and needs.

The cost to the Public Purchase will be minimal. Building Societies, banks, landlords and even customers and their legal advisers will police the scheme by noticing that an SAP is absent - in contravention of the law. The already established position of the SAP in Building Regulations will be compatible with the intentions of this proposal. Consistency will have been established between buildings and motor-cars. (Hopefully, the existing rating declaration requirement for new car sales will be extended to car hire arrangements, thus tidying the situation from both sides.)

Only two requirements will have to be accepted by the EEO and its political masters: First, there must be no attempt, at this stage, to establish evaluative values. The intention is merely to give reliable information, leaving the purchaser or hirer to make a free choice. Markets work when there is reliable information. Second, there must be a D.I.Y. (self-assessment) capability. There is plenty of existing legislation and legal precedent to cope with that, under such concepts as "competent person".

The Energy Saving Trust is the ideal body to administer such an arrangement, in the absence of a return to public administration answerable to democratic control.

9.5.8 To return to what exists; one of the main differences between the NHERS and Starpoint schemes was that the NHERS method emphasised energy costs and included the type of fuel used as a major part of the calculation. Starpoint was less loaded in this way. As a result, the Gas Industry favoured the NHERS approach, whilst other fuels were more favourably disposed to Starpoint. Rival fuel research bodies were used to try to influence the development of both schemes and the situation demonstrated the need for an official, government-backed approach. The SAP has improved the matter but the lesson should not be lost on anyone examining energy conservation administration: The "politics" are not always "political" in the accepted sense. They are more likely to have commercial roots. Energy is big-business. Energy conservation is a threat to all energy markets.

9.5.9 Energy rating is equally applicable to non-domestic buildings but, some might claim, it is more technically complex. The false-start marketing problems of the domestic scene have, so far, been avoided for commercial, industrial and administrative (Tertiary Sector) buildings. The Building Research Establishment has developed what could be regarded as "official" methodology, in the form of BREEAM (BRE Environmental Assessment Methodology), which assesses the "greenness" of a building. At the same time CIBSE has been working on its Energy Code, Part 2 of which provides a method for comparing the energy efficiency of design options at an early stage in the design process. BRE has itself funded the development of a variant to Part 2, which enables it to be used for comparing air-conditioned design options - a major area of potential energy in efficiency. This variant is called TANDEM.

The knowledge is therefore in place for a robust and effective legal requirement for the energy rating of non-domestic buildings, similar in approach to that suggested for dwellings. The non-domestic sector is less encumbered with the mistakes of energy conservation history but does involve the problems of professional sensibilities. However, those who commission or hire commercial and industrial buildings need to be able to assess the "energy value" of what they are buying or hiring. As for dwellings, reliable information can create free markets, which, in turn should lead to energy efficiency and environmental protection.

10 ENERGY CONSERVATION FOR INDUSTRY AND COMMERCE

10.1 GENERAL OFFICIAL ATTITUDES

10.1.1 From the earliest "Save It" times energy conservation 'gurus' have been saying, "We know what to do. The problem is getting people to do it!" This is especially true of energy conservation in industry, commerce and public administration, where managers and directors seem to remain reluctant to embark on energy conservation programmes or to invest to cut energy costs. On the other hand, there is some sense in the views of monetarist politicians who are reluctant to expend public money and public service effort on persuading business people to do what is manifestly to their economic advantage. Energy efficiency and fuel conservation can generally be achieved in a cost-effective manner and make significant contributions to, "the Bottom Line". Conversely, the same could be said of smoking, taking alcohol and over-eating. People know it is to their advantage to stop it but they persist in doing it. Even now that the harmful effects of fuel consumption have been demonstrated, harming the general environment and damaging the health of individuals, attitudes have only marginally improved. Besides knowing what to do, we now know more about why we should do it, but it is still not done.

10.1.2 The situation is not entirely hopeless. Much has been achieved by individual companies and business people, often giving public example and leadership to others. Many professional organisations have taken action; for example, the Confederation of British History (CBI) had an Energy Conservation Working Party that was active for many years (now, unfortunately, less active), the Construction Industry Council (CIC) has an Energy Group, with salaried staff in support, and the Royal Institute of British Architects (RIBA) has an Environmental and Energy Committee. Many local authorities, both institutionally and through dedicated individual employees, sustain local Energy Management Groups. The Institute of Energy, even though a supply-industries-led body, has been active in energy conservation; even to the extent of virtually bailing-out one of the Government's 'orphan child' initiatives, CREATE (The Centre for Research Education and Training in Energy). However, although the Institute of Management (Then the British Institute of Management (BIM)) took over the Dept. of Energy's "Energy Management Workshops" in 1980, the activity gradually faded away and the Institute has been conspicuously inactive in energy management affairs. Therein, perhaps, lies the key to the overall failure of U.K. industry and commerce to be the leaders in energy conservation; in spite of having a demonstrably excellent technical and engineering base on which to build.

10.1.3 As far as ideas are concerned, the U.K. has been prolific in energy conservation for industry and commerce. The problem has been that, since 1979, government has been in the hands of those who are instinctively hostile to subsidy, interventionism and legislation and only too eager to shelter behind the concept of self-regulation. In spite of this, what has been achieved has been through government intervention. Even the Making a Corporate Commitment Campaign (MACC), which involves industry leaders in working out their own energy salvation, was a government-inspired initiative. It did not come from the industrial leaders themselves - even though, at the time, they had their Advisory Committee on Business and the Environment (ACBE). That body was involved only after Officials had created the scheme. And, of course, ACBE itself was a creature of government, not a spontaneous upwelling from "the market". Perhaps energy conservation in this sector has benefited from the presence of a self-confessed "interventionist" in the old Dept. of Energy and now at the Department of Trade and Industry. It is, after all, D.T.I. that is funding the Energy Design Advice Scheme (EDAS) run by the the Dept. of the Environment's BRECSU. This is certainly an interventionist measure and a welcome slice of public purse subsidy in the cause of creating low-energy buildings for industry and commerce and a welcome re-inforcement to the new energy conservation elements of recent Building Regulation legislation. Fortunately, it is not the only subsidising, intervening and legislating activity for energy conservation that has emerged, in spite of government assertions of contrary philosophy.

10.2 ENERGY DESIGN ADVICE SCHEME (EDAS)

10.2.1 The Energy Design Advice Scheme is one of the most effective Government initiatives of recent years, is very strongly funded but has also demonstrated that energy advice, as a commercial activity, even at the highest professional level, cannot survive under normal market conditions - if ever "normal" can be applied to a market. Even otherwise successful and competent business-people will not pay the true cost of energy efficiency advice and consultancy. In the early Eighties, the germ of the present Energy Design Advice Scheme (EDAS) began to emerge from discussions between ETSU and what was then the Department of Energy. The idea was to make expert consultancy in energy efficient building design available to architects and developers at a subsidised rate. This was not only to stimulate the market for energy efficiency design consultants but also to stimulate wider use of knowledge that was being developed by the Government's R&D programmes. The earliest internal document on the idea begins, "Nobody really understands how new ideas diffuse into architectural practice." The scheme, courageously, did not wait for understanding but proposed money-supported action. The proponents of the initiative were talking about a Government investment of some £5 million over five years, which would result in fuel savings of around £140 million by the end of the Century. It was an attractive proposition but it was not widely thought to be likely to be adopted because no amount of "presentation" could disguise the fact that it was a subsidy and a direct intervention in the market. Strangely enough, the Design Support Scheme (DSS), as it was called at this early stage, had attracted sufficient official support by 1986 that it was decided to fund two feasibility studies. During 1986/87 these were carried out by PE Consultants, working with the Building Design Partnership, and by Energy Conscious Design (ECD). At about the same time the University of Strathclyde was seeking support for a similar, but limited, low-energy design support service for Scotland. This became known as the Energy Design Advisory Service (EDAS), attracting funding from the Department of Energy and the Scottish Development Agency and being administered by the Royal Incorporation of Architects in Scotland (RIAS). It is interesting that the Service's first brochure was worded thus, "It enables Architects, Design Teams and their clients to obtain *free advice* on energy issues, and access to expert energy consultancy at *a fraction of the full cost*" (Author's italics). It merely illustrates how full of surprises life is that such a scheme could emerge from a religiously non-interventionist administration.

10.2.2 There was always some controversy about which was the more comprehensive scheme. There was understandable rivalry between ETSU and BRECSU and a wholly unexpected hostility from what was then known as the RIBA Energy Group. The members of the latter were, perhaps, alarmed and/or piqued by what was a blatant interference in the market for energy design consultancy. By 1989 the various participants and interested parties had been reconciled to the formation of a single, U.K.-wide service, which became known as the Design Advice Scheme. After much discussion, it was finally launched in 1993 as the Energy Design Advice Scheme (EDAS), retaining the old Scottish acronym and incorporating the best practices of the previous schemes and feasibility studies. By that time the service was sponsored by the Department of Trade and Industry and had come to involve (Continuing Professional Development) training initiatives in the subject. Administration is carried out through regional centres, based on universities in Northern Ireland,

Edinburgh, Sheffield and London. Subject to availability of funds, the Scheme helps projects that have a floor-area of at least 500 square metres and where there is agreement that the building can be monitored and the effects published. It provides for the costs of energy efficiency design consultancy. (For decades, there have been "energy guru" members of RIBA and RIAS who have given advice to those seeking to build or purchase a low-energy building that comes close to the EDAS concept. They have consistently advised that, having appointed an architect to produce the building, the client should then employ a known "low-energy" architect to monitor the other's designs!) The DTI publishes a quarterly report on the Scheme, entitled "Communique", which contains examples of projects that have been assisted. It has certainly been successful in stimulating the market for low-energy building design consultancy. The European Community has commissioned a further feasibility study, to assess the viability of expanding the ideas into a cross-Europe service. If successful, we should look out for EuroDAS centres by 1996. The scheme is nationally co-ordinated from, EDAS, Camargue House, Wellington Road, Cheltenham, GL52 2AG. Tel: 01242 577277 (Fax: 01242 527277).

10.3 THE ENERGY MANAGEMENT ASSISTANCE SCHEME (EMAS)

10.3.1 During the Oil Crises of the Sixties and Seventies, when the need to "Save It" was unquestioned and when official energy supply demand projections produced the curiously titled, "Policy Gap", a number of measures were introduced to help industry and commerce to cut energy costs. Among these were the Industrial Energy Thrift Scheme (IETS), the Energy Quick Advice Service (EQAS), the Energy Survey Scheme (EES) and the Energy Conservation Scheme. To illustrate the scale of this official response, by the time the Energy Conservation Scheme was closed down, in 1980, it had disbursed some £25 million, EQAS had met over 4,000 calls from industry and ESS had subsidised over 12,000 industrial one-day energy surveys. In addition, there were demonstration project grants, tax allowances and regional development grants for industrial insulation projects. The "Save It" period also created the government-funded Energy Managers Clubs, which were instrumental in creating the National Energy Managers' Courses.

10.3.2 During the Eighties these many subsidies were phased-out - even the meagre £200 per-club-per-year for the Energy Managers Clubs - leaving industry and commerce with little more than exhortation from Government. However, the information and advice services - mostly from BRECSU and ETSU - remained and developed. Today they represent the most competent and professional element of what remains of U.K. energy conservation policy in the form of "The Best Practice Programme", which can still provide some cash, so long as it is never referred to as "subsidy". However, there has been one innovation that, like EDAS, is overtly a subsidy. This was the EMAS. By 1990 common sense began to prevail and the interventionist forces that had introduced the 'Save It' schemes for industry were manifested in the introduction of a scheme to subsidise energy consultancy services for smaller businesses. EMAS (Energy Management Assistance Scheme) was launched in 1992 and might be regarded as elements of the pre-1979 schemes dressed up in 'market forces' clothes. Its failings stem from elements that seek client contribution and try to avoid overt subsidy but it has been, like EDAS, one of the more useful measures of energy conservation policies for industry and commerce.

10.3.3 In essence, EMAS would meet 30% of a consultant's fees to identify energy saving opportunities, 50% of the cost of design advice and 70% of the cost of project management once an investment decision had been made. (For smaller firms, 50 employees or fewer, the 30% identification subsidy could be increased to 50%). There would be no assistance towards capital costs. There was a maximum limit of £25,000 for the largest firms down to a maximum of £5,000 for those with fewer than 1000 employees. An internal report to EEO on the Scheme at the end of 1994 was highly critical of its failure to reach its targets but opinion from industry itself suggests that the terms were not really attractive. Where smaller firms were concerned, there seemed to be some curious practices to improve the attractiveness, which merely reinforced views that energy consultancy is not highly valued and that the scheme was insufficiently generous.

10.3.4 In 1993 the Energy Saving Trust was trying to devise a scheme to help small businesses to improve their energy efficiency. At a meeting convened at the Guildhall in London a proposal was outlined. Observers could not help being amused by the EEO civil servant who, having listened to the proposals, announced that EST had re-invented EMAS but on a more generous basis! This at least indicated that a scheme such as EMAS - or even its pre-1979 'Save It' versions - are the right way to go; so long as they are adequately funded. A corrollory to that is that energy consultancy is difficult to market on a commercial basis whilst energy prices remain low and whilst energy costs are a small proportion of overall costs for most businesses - especially in commerce. (That energy is one of the few and biggest truly controllable overheads remains outside most managers' understanding.)

10.3.5 By 1995 it was clear that Government itself was recognising that EMAS was either not reaching the original target market, that the target market was probably not the one to aim at, that the Scheme was saving little energy and that there was a certain level of mis-administration - to say the least. Faced with a shrinking EEO section that owned it, it is not surprising that, at the time of writing, the EEO is rumoured to have a remit to "privatise" the Scheme. Few in the industrial and commercial side of the "Energy Efficiency Mafia" would be astonished to find such an agency contract in operation by the end of 1995. Many more would not be surprised to find the Scheme closed down, giving another hollow ring to "Climate Change" assertions. (It was closed down in 1995).

10.4 THE GENERAL INDUSTRY AND COMMERCE PROGRAMME

10.4.1 Although so many Government-funded assistance and encouragement schemes for industrial, commercial and public administration energy conservation were terminated after 1979, the sector has remained resilient to total neglect, perhaps because even governments have to listen to informed leaders of the sector, many of whom have set a fine example in energy efficiency. For example, through years of Government neglect of the subject, the CBI maintained an active Energy Efficiency Sub-Committee of its Energy Committee. On the other hand, the Making a Corporate Commitment Campaign (MACC) was a curiously engendered initiative, having been formulated at the time of the industry-heavy ACBE but not having come from it. It is covered elsewhere and is a commendable and major element of the Government's energy efficiency programme - for all its missed opportunities. There are others.

10.4.2 THE BEST PRACTICE PROGRAMME

The Best practice Programme has been a valuable survivor of the general run-down and represents a quality service. Although it does cover some domestic areas, it is predominantly aimed at industry, commerce and public administration. It is run by the Building Research Establishment Energy Conservation Support Unit (BRECSU) and represents a very high standard of presentation and organisation. The scope of the Best Practice Programme is wide. It is primarily an information and advice service but also has some grant support elements, to encourage research and development in innovative energy-saving techniques. For any company or organisation setting about an energy efficiency programme - either for the first time or to renew existing commitment - the Best Practice Programme is an essential first stop. It is similarly useful as a starting point for students of energy conservation politics wanting to get an authoritively-based grasp of the technologies of the subject. It also illustrates what the best of U.K. technology and engineering can produce, given a little funding and encouragement. Its publication-list alone is impressive and the products are authoritative and practical. They are free and there must be few industrial processes or commercial activities that are not covered; from high tech. factory case studies to a programme for energy efficiency in public houses.

The first contact for this service could be either through a local REEO or direct to BRECSU (01923-894040) or ETSU (01235-436747). In 1994 a diskette was produced, "EEO Best Practice Programme", covering all publications relating to low-energy buildings. It can be operated on any PC with DOS or WINDOWS format; again from REEOs or BRECSU.

10.4.3 E.E.O. ENERGY ADVICE DISSEMINATION PROGRAMMES

Mainly under the direction of BRECSU and ETSU, the Energy Efficiency Office has continued a long tradition of conferences, seminars, courses and semi-social events to disseminate energy efficiency best practice and to encourage companies to embark on or renew energy efficiency programmes. It is, of course, an up-hill task to maintain momentum and interest; not helped by the Government's ill-advised ham-stringing of the Energy Managers Groups and the destruction of NEMAC (chapter 8) but, from time to time, interesting new ideas appear.

10.4.4 When Peter Walker took over as Secretary of State for Energy, it was, perhaps, not surprising that his personal enterprising approach should produce his famous, "Breakfast Specials". These events consisted of locally organised "working breakfasts", to which top managers received personal invitations from the Minister. The idea certainly worked in getting high-level business people along to hear the energy efficiency message. As the EEO Director General at the time explained, the success of each event could be judged by the number of Rolls Royces and Jaguars in the hotel car-park! Some guests were, no doubt, chagrined to find dozens of other local businessmen at this exclusive, personal breakfast with the minister but it did get the numbers and exposed them to exhortation, encouragement, exemplar speeches by fellow business-people and exhibitions of energy-saving equipment and services.

10.4.5 More recently, the EEO's "Road Shows" have continued the work, indicating how well leaders of industry and commerce will respond to a clear signal by Government, indicated by ministerial involvement. Both "Breakfast Specials" and "Road Shows", however, did not entirely represent government investment, being heavily funded by the commercial departments of fuel supply industries. By the early Nineties, as the realities of fuel industry privatisation were revealed, this form of funding was drying-up; the trend being set when British Gas "tip-toed away" by announcing a new approach to such co-operation by agreeing to match, pound-for-pound, what the rest of the private sector put into the "Road Shows", rather than giving the usual fixed contribution. Their resulting commitment became a good, round figure - zero.

10.4.6 At a less spectacular level, the fuel suppliers and energy efficiency equipment and services industries have continued to share costs with Government for seminars and courses on specific energy-saving technical issues; some free and others fee-paying. Even the Oil Industry has co-operated in this sort of activity. The programmes are long-established, on-going, practical and innovative. Details are available from REEOs, BRECSU, ETSU and from the EEO itself. In addition to such promotional activities the EEO continues to publish "Energy Management" as a free-circulation magazine for industry and commerce. It has had a chequered career in recent years, as attempts were made to "privatise" it, but it has survived and contains excellent informative articles on energy conservation, in addition to coverage of ministers giving good news on the subject.

10.5 MAKING A CORPORATE COMMITMENT CAMPAIGN (MACC)

10.5.1 The MACC is covered fully in the section dealing with energy accredition (Section 11) because it does appear to stray into that area but, in reality, it is more among "tambourine" initiatives rather than among the rigours of anything concerned with creditable assessment and accreditation. None the less, it has been a successful gesture, has enabled some companies to gain recognition for their genuine commitment to low-energy operation and has shamed others into examining their own position. Its place in the general politics of energy conservation, or as a continuing force for improvement in the energy efficiency of industry and commerce, is likely to diminish as objectives become obscured and mixed with other agendas.

10.5.2 The early history and objectives of MACC are covered in Section 11. Briefly, it was launched in 1992 by the Secretary of State for Energy, with endorsement by the CBI, IOD and ACE. Most of the companies with representation on the Advisory Committee on Business and the Environment (ACBE) became founder Signatories. It was directed at the very top management of companies, at Chairmen and Chief Executives. With conspicuous personal backing by Ministers, it certainly hit that target audience successfully. Within a year there were over 1,250 signatories. The Institute of Energy had also joined the endorsing organisations, which must have been some relief for Government because the Institute had been ploughing its own furrow with its Energy Efficiency Accreditation Scheme. This was and is, a far more professional and rigorous scheme than MACC and had come into being without any EEO backing - at least, not until it was finally established and its backers had developed it in spite of conspicuously luke-warm reactions from Officials.

10.5.3 MACC itself was always going to be a difficult scheme in terms of maintaining momentum because its lack of rigour meant that nothing much happened to nor was demanded of signatories after their initial involvement. In December 1994, at a meeting that was presented as an AGM of Signatories, there was a marked absence of Chairmen or Chief Executives. In addition, the Secretary of State's address linked the scheme to a new initiative, the Environmental Technology Best Practice Programme, to which MACC participants had some sort of privileged entree. However, this development seemed most likely to dilute what energy efficiency thrust remained in MACC and to confuse the original messages with those of environmental protection and non-energy waste management. Although interest in environmental protection has re-kindled interest in energy conservation, any government is going to have problems in pursuing environment, pollution and energy conservation collectively without losing focus on any of them. The French re-modelled their old Energy Management Agency into an environment and energy organisation but, in the U.K., although the old EEO has been partly absorbed into the Department of the Environment, it is far from integrated. The excellent energy Best Practice Programme at BRECSU is now both EEO and D.T.I. funded. This new Environmental Technology Best Practice Programme is announced as, "A joint DTI and DOE initiative managed by AEA Technology through ETSU and the National Environmental Technology Centre". It will be interesting to see whether this has been a wise use of the term "Best Practice Programme". Will the new initiative benefit from the brand images of the established scheme or will it confuse and dilute? Making it a centre of focus at the 1994 MACC AGM was a surprise that bemused. In any event, MACC remains, was a brilliant concept but it remains to be seen whether the EEO will be allowed the resources and clout to maintain it as a credible commitment.

10.6 A WAY AHEAD

For the future, energy conservation in industry and commerce - including public administration; (what main-land Europe generally terms, "the tertiary sector") - may well benefit from a change of emphasis in the future development of the National Energy Foundation. It is generally understood that its ambitious plans for the "Oddessy Project" - an Eighties-style mega energy centre - have been radically altered, scaled-down and are to be re-focused to concentrate above the domestic sector. A national energy centre run by NEF for industry and commerce could build successfully on both government and non-government initiatives and on the extensive goodwill and enthusiasm in the sector that has survived the general trend of neglect.

11 ENERGY EFFICIENCY QUALIFICATIONS AND ACCREDITATION

There has been a long-standing recognition that it is sensible to have some sort of training in energy management. From that arose the notion that some sort of qualification would be desirable. In more recent years the idea of recognising the energy efficiency of individual firms and organisations has gained acceptance. Training and qualification for individuals has been haphazard and unsatisfactory. Recognition of energy efficiency for organisations has progressed but with the usual muddle and confusion between "official" and "voluntary" that one has come to expect in the U.K. Qualification for the individual and accreditation of the company are distinct issues but they probably share a lot of common ground and could enjoy complementary development - given the least political will.

11.1 TRAINING AND QUALIFICATIONS FOR ENERGY MANAGERS

11.1.1 At undergraduate and post-graduate levels there are dozens of courses for energy management on offer in the U.K. Many disciplines include energy studies in their curriculum. The main problem is that, in general, such courses are either too technology biased or deal too much in global energy generalities. Neither bias can be expected to turn out an "Energy Manager".

11.1.2 In the late Seventies, Government introduced a one-week course for newly appointed energy managers. It recognised that an effective energy manager could come from one of many disciplines, from accountancy to fuel engineering, by covering subjects ranging from energy audit to motivation and including accounting, targeting and monitoring. It certainly looked as though the importance of "management", as opposed to technology, had been recognised when, in 1980, these courses were handed over to the British Institute of Management and marketed as, "The Energy Managers Workshop". However, they gradually became more and more technology based and, at the same time, suffered from the general decline of demand for industrial and commercial training. Without direct encouragement from government, this initiative has quietly faded from the scene. Other courses have appeared and disappeared and some very novel ideas have been tried but nothing seems to have become well established.

11.1.3 In the Eighties a group emerged called the National Energy Efficiency Association (NEEA), which sought to establish professional recognition for energy management. It gained some private support but, inevitably, lacking direct government encouragement, it failed to become established. The NEEA appeared at about the same time as the National Energy Managers Advisory Council (NEMAC) was wound up. It was finally launched at the London HEVAC Show in 1989 as, "The Professional Association for Managers of Energy". It began to produce a journal, "Energy Report", but the last edition appeared in September 1991 and, along with the NEEA itself, has since been in hibernation. This in no way reflects badly upon the originators nor upon the members. Their initiative happened to co-incide with the worst period of the Second Thatcher Recession, when industrial and commercial sponsorship became very hard to obtain. Like the Energy Managers Groups, NEEA needed the most trivial sums to survive. Unhappily, the personal enthusiasm of its Chair and Members was insufficient without even such modest financial backing. An insignificant drop from the EEO budget would have given the U.K. a world-leading development in the professionalising of energy management. A mere morsel of intervention by the President of the Board of Trade - perhaps before a coffee-break, if not before "breakfast, lunch and dinner" - would have done the trick. Although in hibernation, NEEA still exists, with about 50 loyal members and plans to re-launch in 1995/96. Contact may be made through P.O. Box 526, Ingatestone, Essex, CM4 9TP.

11.1.4 It is just posible that the apparent failure of government to encourage or support such efforts as those of NEEA may have had much more to do with Whitehall than with Westminster. There has been an enormous investment of resources, reputations and career prospects in the NVQ/SVQ movement (National Vocational Qualifications and Scottish Vocational Qualifications); in what might be interpreted as a desperate attempt to bring the UK nearer to the technical competence of its EU partners without the intellectual rigour and social investment that has purchased such competence.

11.1.5 Whilst NEEA and others were attempting to establish courses, examinations and qualifications for energy management, various Whitehall departments were ploughing their own furrow in the NVQ/SVQ fields, with what now appears to have been perverse disregard for the efforts of the working energy management enthusiasts. With joint funding from the EEO, British Gas and the Electricity Association, the EEO itself was producing the EEO/MCI Standards for Managing Energy. The Department of Employment was involved, along with BRECSU and

ETSU. The product is managed by MCI (the Management Charter Initiative.), an employer-led organisation that is "backed by government" and set up to develop and establish National Standards of Best Practice for UK managers and supervisors. The project has the following aims:

- to set out the skills, competence and functional roles in energy management;
- to produce a set of generic Standards for Managing Energy to meet the needs of supervisory, first-line and middle managers responsible for the use of energy resources;
- to encourage sector bodies to produce sector guidelines;
- to promote the adoption of energy efficiency issues in NVQs and SVQs.

11.1.6 In general, these aims correspond to the sort of things that the NEEA and others have been advocating, for some time, as a frame-work for energy management training and qualification. The problem - as with much of the NVQ/SVQ activity - is that the initiative appears to be all about making awards on the basis of "portfolio evidence" rather than with producing rigorous training and certification. It will recognise what exists but may not produce new effort. In fact, it all sounds suspiciously close to the advertisements one sees from certain overseas universities, offering degrees without study or examination - " recognising the University of Life" ! Of course, if the assessments in the process are strict, they will produce a need for training, from which may emerge the long-overdue, real, academically worthwhile courses for energy managers. If they are not, then the process will do little to enhance the quality of UK energy managers and supervisors. It will merely produce yet more sets of worthless letters. It should be surprising that employers' organisations are accepting this initiative, " backed by Government". (In all fairness, it should be noted that the NEEA proposals themselves were not entirely devoid of "portfolio evidence" aspects, with some members not entirely happy about certification based on pass/fail examination.)

11.1.7 In the absence of progress by NEEA and with the lack of academic credibility of the EEO/NCI project, the U.K. can offer no specific energy management qualifications. The various under-graduate and post-graduate courses offer only generalities in comparison to specific needs. If Government were able to reverse the ghastly mistake that was made when Energy Management Groups and NEMAC funding was withdrawn, that movement could well be the basis upon which NEEA and professionalisation of energy management could be grown. However, like energy accreditation of companies (see below), it will also be necessary to avoid the usual British fudge and to ensure professional and academic rigour - perhaps even examinations! - which seems to attract so little support in the U.K..

11.2 ENERGY ACCREDITATION

11.2.1 Making a Corporate Commitment.

The matter of energy accreditation of companies has been a curious mixture of the official and the private; curious because the impetus for rigour and value in a scheme has come from industry itself whilst the Government's own initiative is largely weakened in credibility because of official reluctance to impose any policing on the industrial participants. In reality, the door was already opening when Government first launched its, "Making a Corporate Commitment" (MACC) Scheme and a little courage by ministers and officials could have given the U.K. a lead in this matter on the international stage. Instead, a largely P.R. scheme emerged which enabled the usual players to strut upon the energy efficiency stage without too much effort. What is such a pity is that many firms - and their individual Chairmen and Chief-Executives - did make a genuine commitment to improve their company's energy saving performance and would have stood-up well to scrutiny of their claims. They would have set a fine example, gained well-deserved public acclaim and would have given a lead to others. Instead, in spite of hints and rumours of action, successive ministers accepted the signed commitments, publicised them in well-produced glossy brochures, organised many excellent exhortation events but stopped short of real scrutiny. In 1992 the Parliamentary Select Committee on the Environment criticised the Government for failure to monitor and publish the energy performance of companies signing up to the commitment. That Committee was chaired by the present "Minister for Energy Efficiency"! It was left to an independent initiative to give those who deserved it an accolade that had any worth.

The MACC Scheme itself is an excellent idea, well produced and promoted. It was launched by the Secretary of State for Energy, John Wakeham, in 1992, with endorsement by the Confederation of British Industry (CBI), The Institute of Directors (IOD) and the Association for the Conservation of Energy (ACE). Most of the companies with representation on the Advisory Committee on Business and the Environment (ACBE) became the Founder Signatories. In fact, the Government was able to display an impressive 34 top companies as its Founder Signatories, including some fuel suppliers but, conspicuously, omitting any of the newly privatised electricity suppliers or producers.

Each company made a seven-point declaration of commitment, which would be displayed in its premises. The certificate was useful both for external P.R. purposes and for encouraging the company's own staff. The seven commitments were:

- Publish a Corporate Policy
- Establish an Energy Management Responsibility Structure.
- Hold Regular Reviews
- Set Performance Improvement Targets
- Monitor and Evaluate Performance Levels
- Report Performance Changes and Improvements to Employees and Shareholders

Within a year, the Secretary of State for the Environment (The Department of Energy, that launched the Scheme, had now been disbanded) was able to announce that over 1,250 organisations had signed-up to the MACC campaign. The Institute of Energy had also come aboard to endorse the initiative.

The appearance of the Institute of Energy endorsing MACC was interesting because it had been pursuing its own more rigorous form of accreditation for some time, with the Energy Systems Trade Association (ESTA). This became known as the Energy Efficiency Accreditation Scheme.

11.2.2 THE ENERGY EFFICIENCY ACCREDITATION SCHEME

The ESTA and Institute of Energy scheme had been in the making for many years, before it was finally launched at the 1993 NEMEX. Both organisations fought a long but successful battle in seeking funding from industry and commerce.

The Scheme gives industry an opportunity for achievement in energy efficiency to be independently assessed. Unlike MACC, which is merely a commitment to do better, the Energy Efficiency Accreditation Scheme examines what a company is achieving and gives accreditation only where it is deserved. It costs a company £2,500 in fees to become accredited and there is a requirement for renewal examination. It could be said that the Scheme is altogether very un-British and, as such, does represent a major break-through for energy efficiency in the U.K. Unlike BS quality schemes, the Accreditation Scheme expects performance and achievements in addition to procedures and documentation. A year after its launch only 5 of the Founder Signatory companies to the Government's MACC had become Accredited Companies and it will be interesting to see how this develops or whether the Energy Efficiency Office, which claims to support the Scheme, will use it to check-up on the companies that have signed-up to MACC. The accreditation Scheme itself acknowledges that participation in MACC would be an element of evidence that a company has a coherent energy efficiency policy. A reciprocal requirement would go a long way towards giving MACC some real value and would strengthen the Accreditation Scheme itself.

Companies wishing to participate in the Scheme may obtain details from ESTA (PO Box 16, Stroud, Gloucestershire. GL6 9YB - Tel: 01453 886776).

11.2.3 There has certainly been some progress in the U.K. in enabling companies to gain energy efficiency accreditation but the prospect remains unsatisfactory for individuals. As far as companies are concerned, it would be a further step if there were a requirement to have an energy efficiency statement in annual accounts. Both the CBI Energy Efficiency Working Group and the first session of ACBE made such a recommendation. It had even appeared, in the Seventies, in ACEC documents. As environmental reporting gains acceptance and prominence, energy efficiency accreditation and statement may gain ground as part of that reporting.

Legislation, requiring companies to have certificated energy managers in control of their energy use, would, of course, solve the problem of professionalising energy management in the U.K.. Although this has been done in Japan, it is unlikely to be on the agenda here. Alternatively, official encouragement of a recognised, rigorous and credible energy management diploma would introduce some order into a chaotic situation. This should not threaten the existing providers. On the contrary, a national norm and curriculum for personal certification would give them a recognised training to provide and, probably, enhance the demand. Meanwhile, the prospect for the individual in this area remains bleak.

THE EDUCATION SECTOR

12.1 ENERGY CONSERVATION IN THE SCHOOL CURRICULUM

12.1.1 At first sight, the issue of energy conservation in education is simple; as encapsulated in an introductory note to a report by the Martin Centre for Architectural and Urban Studies in 1988: "Government in Britain has aspirations that children should be encouraged through their formal schooling to adopt pro-conservation attitudes and behaviour". However, in practice the issue is far more complex. First, there is fierce competition for space in the school curriculum, at all levels, and the curriculum itself presents a dynamic, ever-progressing situation. Secondly, there is disagreement about the place of energy efficiency among curriculum subjects: Does it belong to one or several selected subjects or is it "cross curriculum" - or both? Third, there is confusion about what "energy efficiency" is in the curriculum context, whether it is part of more general "energy studies" or whether such a term itself implies teaching energy efficiency as an end of such study. Fourth, it is not entirely agreed whether it is educationally ethical to take a prescriptive approach: not merely as a philosophical issue but also because promoting an "answer" to energy efficiency can lead into fuel-choice, equipment-choice and life-style choice areas - where vested commercial and social interests are only too anxious to be involved. Finally, there is the issue of energy conservation in schools as structures and the obvious desirability of linking that to the energy efficiency teaching that occurs in those structures. This raises both opportunities and tensions; for example, between the Government's Best Practice Programme and the Government's curriculum development policies. A conference at BRECSU in the Spring of 1994 explored this issue but it is not yet clear how much funding government itself is prepared to put behind development, to produce a policy, programme and resource material to combine the pedogic aims with the school management requirements.

12.1.2 In the Seventies, arising from the "Save It" imperative, and in an era of more inventionist views in Government, there was much activity at official level and the then Energy Conservation Unit of the Department of Energy had a well staffed section devoted to the matter. By the early Eighties activity was reduced, just at a time when the educationalists themselves were becoming more aware of the subject and the need for action. Individual university departments of education had recognised the need. They were producing useful and novel work, many working with progressive local authority education departments. Unfortunately, this progress was not matched by resources within the Department of Energy; so that, by the end of the decade, there was little Energy Efficiency Office involvement. It was not until the early Nineties that the first signs appeared of renewed official interest.

12.1.3 During 1985 the EEO commissioned a report, "Energy Efficiency Studies and the Secondary School Curriculum", which, within an overall action-plan, had concluded, "There is apparently little resistance to the idea of energy efficiency studies, only a lack of precision about what it involves and a severe shortage of appropriate resources". This was, of course, referring to teaching resources but it might as well have referred to a shortage of policy-making and management resources within the Energy Efficiency Office for the issue.

As things stand at present, the Energy Efficiency Office itself is not equipped to tackle the issue, the Energy Saving Trust is not financially resourced to tackle it and the burden falls on a body known as CREATE - The Centre for Research, Education and Training in Energy - an acronym-led title and misleading, for a body that was supposed to be concerned with energy efficiency in the school curriculum. The National Energy Foundation, Neighbourhood Energy Action and others are active in this area and all generally have a relationship with CREATE, which, as explained below, at least establishes a principle and potential, if not actually fulfilling the real need.

12.2 CREATE - THE CENTRE FOR RESEARCH, EDUCATION AND TRAINING IN ENERGY

12.2.1 By the early Eighties, many local authority Education Departments had recognised a need for energy conservation to be taught in schools. These, plus individual schools and university faculties of education, had begun to produce teaching resources and to embark on research into the educational issues involved, both practical and theoretical. Most made approaches to the Energy Efficiency Office for endorsement or financial support - or both - and, finding little of either, began to turn to non-government sponsors in industry and commerce. Given the great number of L.A. education authorities and university education faculties who were nearly all becoming active in this field, the potential sponsors began to respond negatively to the level of demand - generally replicated many times over. They, too, turned to the Energy Efficiency Office seeking some sort of official co-ordination. It has to be said that many of the potential private sponsors were willing to help but were simply overwhelmed by the level of demand and were irritated by the re-iteration of the same or over-lapping proposals. The brunt was borne by the fuel suppliers, which were still publicly-owned utilities and already had good reputations for producing educational resource materials. These suppliers had two main reasons to be active in the area: to produce generations of young consumers well-disposed to their particular products and to ensure that they had the pick of the best brains emerging from the education system; being industries needing high levels of technical and scientific expertise. Inevitably, these players were competitive in their approaches to the education sector. That, in itself, is a major consideration in the politics of energy conservation in this field. Unfortunately, much of the investment in energy conservation education by the fuel suppliers has been reduced as the U.K. privatisation drive has blunted inter-fuel competition - which may well have been the real agenda!

12.2.2 In 1983 the Energy Efficiency Office called a number of informal meetings with the fuel supply industries in response to their demands for co-ordination of the growing demands for sponsorship for curriculum development and teaching resources for energy conservation. At one of these a level of acrimony had arisen, chiefly from exasperation by the industries at lack of a Government lead and from the Officials' frustration of having to operate with dwindling resources. An Official had angrily said something like, "If you are all so clever, why don't you do something about it?". One of the fuel suppliers - the then still publicly-owned British Gas - took that as a convenient official request to steal a march over its competitors, where it was already dominant. In consultation with teachers' organisations, L.A. Education departments and some education academics, a first step to co-ordination was devised in the form of a series of conferences to be run during 1984. At the time the Secretary of State for Energy was Peter Walker, whose enthusiasm for the subject of energy conservation was noticeably out of step with the general trend of policy at the time. It was he who had stated that the U.K. was by then lagging behind in energy conservation; so it was not surprising that he gave such strong endorsement to the conferences and wrote into the conferences brochure:

"Real long term energy security for this country will be in the hands of those who are now at school or about to start their education. Unless they are aware of the problems, understand the real issues of energy policy and know how energy management can be applied to their styles of living..... one cannot feel that an energy-efficient society can be assured. .....I look forward to receiving reports from the conference chairmen....."

12.2.3 The conferences, under the title "Energy in the School Curriculum", took the form of one programme repeated at three locations - the University of London, the University of Bristol and the University of Leeds. They were wholly financed by British Gas with the prominent involvement of the Energy Efficiency Office, including high-level official spokesmen. Each was chaired by the local Professor of Education and the three of them duly reported to the secretary of State. Their findings, in summary, were:

- Further conferences were needed, more questions than answers having emerged.
- Local Education Authorities were desperately short of energy conservation teaching resources.
- In-service training in the subject was needed.
- Some sort of working party or liaison group was needed for, "shaping policy for implementation of change."

12.2.4 The conferences certainly stimulated follow-up work. Others were arranged at local authority or university departmental level; the EEO recruited specialists, produced excellent energy efficiency teaching resource packs and commissioned the report, "Energy Efficiency Studies and the Secondary School Curriculum" (1986), mentioned above. The fuel supply industries continued and increased their own output of teaching resource materials but now re-focused by the conference discussions and report.

12.2.5 Early in 1984, whilst the Curriculum Conferences were in progress, a paper had been produced by the Department of Architecture at the University of Cambridge entitled, "Lay Views of Energy Conservation in Britain: The Significant Case of Primary School Teachers". It concluded with a telling comment on a dichotomy between teachers and government officials:

"Because of this divergence in their views about, and their preferences for, conservation, agreement between these two groups about when, if and how it should proceed is likely to prove problematic and elusive. Instead, both groups need to be aware that ..... they seem more likely to talk past each other in mutual disregard and misunderstanding".

In spite of renewed activity, this view continued as a valid reflection on the issue; so that, by 1986, there was a general pressure from many interests for a repeat of the 1984 Curriculum Conferences exercise. Again, British Gas financed the events in the absence of any Government financial contribution; though, it must be admitted that the EEO support "in kind" was much greater for this round and the Under-Secretary of State for Energy attended the first conference, with his speech repeated by the EEO Director General at the other two. The earlier formula, of repeating the same programme at three venues, was again used - chiefly to overcome the difficulty of gathering teachers and educationalist from all over the U.K. Again the Secretary of State for Energy himself called for a report and it was interesting that, in his supporting message, this time he spelt out the official commitment: "My own officials have been closely involved from the earliest planning stage and will be participating at all three conferences." They did! The conferences took place in January 1987 at the Universities of London, Bristol and Manchester. From them came the Chairmen's Report that produced CREATE. (The Centre for Research Education and Training in Energy). Peter Walker received it in person and the three Conference Chairmen, representing the general view of the Profession, very nearly obtained what they wanted but the gains were spoiled by the needs of political dogma.

12.2.6 This 1987 Report re-iterated the 1984 Report, on matters such as teaching resources and in-service training, but its main thrust was in repeating the need for policy co-ordination. This time it came down strongly in favour of a government-directed and government-funded body: "...the Energy Efficiency Office should establish a standing co-ordinating body to carry forward and develop the ideas and convictions that have been expressed at the Conferences..." The Report went on to clarify what such a body should be capable of:?

- giving direction at national level
- co-ordinating work being done at local level:
- initiating in-service training
- initiating projects to establish energy efficiency in pre-service training
- co-ordinating involvement by government departments, education authorities, other organisations and industry
- identifying future needs and developments, initiating research and disseminating results
- relating such work to general public education and adult education, as well as to the school curriculum."

12.2.7 The Report also picked up an important view about any government-directed activity that is taken to encourage social behavioural change: "...such a move would give a clear signal that government is genuinely interested in creating an energy-literate society". Apart from the profound truth of this view, this and other parts of the Report perhaps did much to introduce and establish the concept of "energy literacy" into the energy conservation issue.

Even if the Government was not "genuinely interested", it is much to Peter Walker's credit - as in much of his non-Thatcherite energy conservation activity - that he made a personal, written response to the Report to the three Conference Chairmen. In it he made a specific comment on the recommendation for the body that was to become CREATE: "Your main recommendation, which I intended to follow up, stresses the need for national co-ordination..."

12.2.8 By the end of 1987 the Centre for Research, Education and Training in Energy had been set up but it immediately ran into funding, organisational and directional problems; chiefly, because Government resisted setting up a properly funded, government-directed body. Instead, political correctness demanded a quasi-business/quasi-professional creature that had to raise funds for its continuance by trading in an already established and competitive area but over which it had the unfair advantage of having an "official" aura. It was dominated by its mainly fuel-supply Founder Members and lacked the guidance of the high level educationalists that had informed its originating conferences. It has since soldiered along under these handicaps, has certainly produced some excellent teaching aids but has never been able to fulfil the directional, co-ordinating and policy developing role for which its originators had intended it. It has been one of the most discouraging missed opportunities of the energy conservation scene since 1979. Its very name was also a mistake, subsuming the intentions under the total "Energy" umbrella, instead of making a positive statement about "Energy Conservation" as an end in itself.

12.2.9 The Founder Members were: British Gas, British Nuclear Forum, Esso U.K., the Institute of Energy, Powergen, National Power, Sainsbury's and Yorkshire Electricity. The Energy Efficiency Office was also a Member and provided a share of the start-up funding. However, this was to be withdrawn after 3 years, when the Centre was expected to stand on its own feet, funding itself with earnings from development projects. At once, the other Founder Members began to complain because, having put sums into the Centre, they found themselves being charged for projects in which they had an interest. At the same time, other groups - either straightforward business or non-profit-distributing organisations - saw CREATE as a new, subsidised competitor in developing and selling energy conservation teaching materials. Without consulting the other Founder

Members, the EEO had sought-out and appointed a Director and fixed a level of remuneration that at least raised other Members' eyebrows. Offices were established where staff happened to live (At one time, scattered between North Wales, East Anglia and the Home Counties), without reference to the Management Committee. One of the Founders, British Gas, withdrew funding and left. Above all, as mentioned earlier, the Management Committee did not contain any educational academic expertise of any standing; some of the Founder Members' representatives were just P.R. and marketing people. The set-up was certainly not appropriate for a policy-driving and national co-ordinating body.

12.2.10 In spite of the inadequate policy-level resources, the Centre achieved much at the operational level of teaching-resources and information. An enquiry and teaching-resource data-base was quickly set up and the Centre was contracted to process enquiries on behalf of the EEO - and continues to do so. A register of interested teachers was established. Several teaching material projects were carried out, producing units and packs of very high quality. The Centre recently did work for the Energy Saving Trust, producing a "Directory of Energy Educators". However excellent, the activities were merely the sort of thing any commercial or quasi-commercial educational consultancy might carry out. With the EEO running down its own educational resources and becoming more distant from the Centre, the U.K. still had no "standing co-ordinating body", as the 1984 Conferences Report had sought. Funding became progressively more difficult but credit must go to the Institute of Energy for keeping the show afloat, often by the dedication of individual enthusiastic members.

12.2.11 Into the mid-Nineties the Centre's problems continued. Its management body received continuing reports of difficulty in increasing membership from the private sector, on which Government plans for CREATE had always been based. Furthermore, the EEO decided to fund the Enquiry Service only in so far as enquiries related to energy efficiency; general energy enquiries being regarded as someone else's concern. This is an entirely understandable view and high-lights yet another weakness of the arrangements: A brief that is wider than energy efficiency is required for commercial viability but moves in that direction dilute the Centre's prime purpose and its attractiveness to its Founder Members - demonstrably so in the case of the EEO. In the Summer of 1994 the centre changed the name of its Newsletter to, "Energy Watch"; thus affirming its move into the wider field and raising doubts about its energy efficiency long-term credentials (The student of energy efficiency politics might detect a similarity between this situation and developments in the also "privatised" NEMEX).

In 1994 the Energy Efficiency Office began to re-build its own educational resources. The Centre's Chief Executive was also able to report, "...renewed confidence shown in CREATE by the Energy Efficiency Office". If nothing else, such a statement of renewal illustrated how the EEO had become estranged from its child in the previous years!

12.2.12 In spite of its unpromising history and financially precarious situation, CREATE remains a positive development in the educational sector of energy conservation. It works, or has worked, with the Energy Saving Trust, the Institute of Energy, the Building Research Establishment Energy Conservation Support Unit, the National Energy Foundation and Neighbourhood Energy Action, besides other organisations in the sector. However, this activity has only been at the operational support level. CREATE is not a policy driving body. Only real, government core-financial support can make it that - "the clear signal that government is genuinely interested" - so clearly stated by the 1987 "Energy in the School Curriculum" Conferences Report. The Energy Saving Trust is known to have an interest in promoting energy efficiency through the school curriculum and it would certainly be the obvious "home" for a real CREATE. Similarly, the National Energy Foundation would be as appropriate. Both bodies, like CREATE itself, are powerless through under-funding. It is a depressing but familiar situation. Comparatively minute sums of public money would yield impressive results. Without it CREATE must struggle on but it does at least provide a starting point and centre of focus for energy conservation in schools. The present contact address is : Kenley House, 25, Bridgeman Terrace, WIGAN, WN1 1SY. Telephone: 01942 - 322271, or to the Registered Office at the Institute of Energy, 18, Devonshire Street, LONDON, W1N 2AU.

12.3 INTERGRATING CONSERVATION INTO THE CURRICULUM

12.3.1 As CREATE and others have discovered, teaching energy conservation in schools needs to be clearly defined and distinguished from energy studies in general. Even when acknowledging energy conservation as a subject in its own right, the teaching profession has a propensity to drift towards re-newables in its lesson content. "Tilting at Windmills", might be a punning warning. In the schools sector - and even in higher education - that is the trap. However, there is an encouraging tendency for energy conservation teaching in schools to be linked to practical energy management in the school itself, which will help to keep the issue properly focused. This is being encouraged by EEO and BRECSU activities. An even worse enemy is the pedant who scuppers the whole matter by insisting that energy cannot be conserved and, therefore, "Energy Conservation" as a subject is an intellectual nonsense! In the late Seventies the Association for Liberal Education (ALE) had made valuable contributions to the development of energy conservation in the curriculum and had confronted and rebuked that attitude - chiefly emanating from physicists. It remains an irritating distraction for those who are trying to make real progress in this area.

12.3.2 Energy conservation does involve a wide range of disciplines - even in its practical application - let alone in including it in the school curriculum. Its technicalities include mechanics, physics, chemistry and a whole range of environmental disciplines; even before economics, politics and other sociological subjects are considered. It is little wonder there is a strong demand by teachers for both pre-service and in-service training on the subject. In 1980, in his, "Report of the Advisory Committee on Energy Participation", for the Council of the University of London's Department of Extra-Mural Studies, Harry Frost wrote, "A youthful tendency to favour black-and-white descriptions cannot make for easy teaching when the topic is largely grey". Most teachers and those working to promote energy conservation in the curriculum would appreciate the sentiment. That it came from a study examining the same issue, but for further and higher education, is a reminder that the higher sector presents challenges similar to those encountered in the schools curriculum.

12.3.3 Even if teachers are able to keep energy conservation in focus, among the beguiling clutter of energy issues in general, they are faced with the problem of making room and time for it among an already crowded curriculum. A round-about way of generating teacher interest was tried in the early-Eighites by the EEO by means of competitions for schools. They were notoriously unsuccessful. Even after massive publicity, the prospect of the Prime Minister herself presenting the prizes, plus funding from an international oil company, only 57 entries were received for the EEO's 1985 Schools Energy Competition from a target population of over 8,000 schools. These were not widely publicised figures. The top-level prize-giving went ahead. No-one appeared to have told the P.M. that she was attending a failure but the Under-Secretary of State for energy at the time, Lord Avon, immediately wrote to the Chairman of the British Gas Corporation, asking whether the Company would take-over organising the competition. That a still publicly-owned, nationalised industry should be able to do better in the white-heat of marketing than an international oil company was clearly not then an issue with government!

12.3.4 The approach taken by British Gas was a useful indicator of how the teaching profession could be approached successfully in energy conservation matters. Instead of imposing an overtly Energy conserving competition on the schools, the organisers called meetings of teachers to find out what they wanted. As in the energy in the School Curriculum Conferences, the message was simple and clear: teaching-resources, especially computer software. As a result, high-grade software was developed, that would have applications beyond the competition, and "Energy Study U.K." became a computer-based task with educational objectives that just happened to have an energy conservation content. In addition, the staff of the Company's Schools Advisory Service (now a casualty of "privatisation") individually "sold" participation to teachers with free issue of the computer software as a commercial "come-on". 3,600 schools entered the competition. Teacher enthusiasm even led to development of the computer software; so that it became a marketable product as CEDRIC (Computer Energy Display and Retrival of Information and Calculations). By aiming the competition at the higher-prestige faculties of science, mathematics and computer science the organisers had overcome the "back-to-nature" or "on-with-the-hair-shirt" image that had hitherto been attached to energy conservation. It was a lesson for any future development but it was not universally accepted. The task of integrating successful energy conservation into the school curriculum requires high grade research, expert resources and far more dedicated effort on the part of government - including generous funding. The successful competition indicated what could be achieved. It did not solve the problem of integrating the subject into the curriculum.

12.4 ENERGY CONSERVATION IN FURTHER AND HIGHER EDUCATION

12.4.1 Where energy conservation in further and higher education is concerned, its teaching presents much the same problems and raises much the same issues as it does whithin the school curriculum. These may be summarised as arising from the questions of whether "energy conservation" is to become a discrete discipline, whether even "energy" itself is too narrow to be a discipline or whether both subjects should merely seek to establish themselves within existing disciplines - modules in a modular approach. However, there is also the matter of professional or vocational training, which does not arise within the school curriculum. As energy policy, energy purchasing and energy efficiency have become professions, there has arisen a demand - either implied or expressed - for further and higher education in these subjects. With the emergence of environmental protection and pollution control, the place of energy conservation in tertiary teaching has, at the same time, become more urgent to define and more complex to discuss. Similarly, some already-established professional courses in universities are recognising the need for energy efficiency teaching to be included in their curriculums; for example, in architecture, building services, surveying, urban planning, industrial design and so forth. Where the whole scope of building is concerned the issue has even caused discussion about which profession should take the lead in the matter.

12.4.2 Among what might be termed, "the traditional university disciplines", those that clearly have an interest in energy and energy efficiency are Physics, Mathematics, Chemistry, Geography, Economics and Engineering. Among the "new" disciplines, energy is certain to interest Politics, Environmental Studies, various branches of Social Sciences and Public Administration. However, these faculties are notoriously disinclined to adopt prescriptive approaches to study - a "grown-up" version of the same situation that can arise within the school curriculum - whereas many of the forces trying to establish energy efficiency/fuel conservation in further and higher education come from precisely such a prescriptive position. The problems and antagonisms that arise are unlikely to be solved so long as the U.K. has not solved the underlying problem of reconciling the positions of higher education and vocational training. Those studying the politics of education can only register that such a background exists and that it will inform discussion and policy for as long as may be foreseen. Within universities and other higher education establishments there is likely to be a continuing reluctance to adopt multi-disciplinary approaches, which is likely to be evident also where vocational training is concerned. Within the fashionable "commercialisation" of further and higher education, energy conservation will probably find niche markets for cross-curricula "products" and will be able to to "make a killing" in the scramble for under-graduate and graduate customers. This itself raises a key issue on the subject, 'Who are the true customers for energy conservation courses?'.

12.4.3 Where higher education in its true sense is concerned, it is society itself that must be the customer but that is to move into difficult areas that are not properly the concern of energy conservation. At the most venal level, the student is looking for courses that will bring employability, which is also the concern of employers in industry, commerce and public administration. The difficulty is that the employers have not been able to articulate what they require; as a result, the students are uncertain about what they should be demanding of the educators. At the most basic, technical level it is possible to understand and to define what an Energy Manager should know and in what ways s/he should be coached to exercise judgement within that knowledge. It is less clear whether such knowledge is, indeed, degree level - first or higher - or whether it is merely the content of a diploma that could be obtained with or without a degree. Similarly, it is less clear what sort of academic or training input is required to produce minds that can operate with "energy policy" occupations - whatever that may mean - as opposed to being employed in the more easily definable vocation of "Energy Manager".

12.4.4 The training of architects encapsulates the many issues that confront anyone studying this problem. It is particularly apt because it starts from the position of uncertainty about its own situation within the training-versus-education debate. Is architecture an academic subject or is it merely a particularly sophisticated technology? In very traditional academic terms, Is it Arts or Science? Part of the Profession itself has been in the vanguard of energy conservation in the U.K. for many decades but the majority remains resolutely aloof from such technicalisation of, "the Senior Profession". When environmental issues arrived on the national agenda, it was interesting to see how a whole new constituency of architects and architect academics became involved but still seemed uncomfortable with the established low-energy agenda. However, many Schools of Architecture have, on the whole, become powerfully concerned with low-energy building issues. COHSA (The Committee of Heads of Schools of Architecture) takes an active interest in the RIBA Environment and Energy Group and has an active minority of its members prominent in energy conservation initiatives. However, as in the Profession itself, the many excellent initiatives are the work of a minority. The Environment and Energy Group is not "mainstream" and even the language of energy efficiency is far from being current among architect undergraduates and their teachers.

12.4.5 As things stand at present in the U.K., it is difficult for Government to have much influence on what is taught in higher education. Where architect training is concerned, Government is most concerned with reducing the length of the training and, therefore, unlikely to wish to see yet another subject included in the architectural curriculum. Moreover, Government would see it as the responsibility of industry and commerce to be the driving force behind curriculum content. There is even a view that Government would gladly see architecture "de-professionalised" and tackled as some kind of adjuct to more general and technical training in building structure matter. Whatever the arguments for or against that, it could well make it easier for energy conservation to make an impact on course content. The existence of the recently-formed Energy Committee of the Construction Industry Council (CIC) could offer an already combined approached route for the many disciplines of the construction industry.

12.4.6 Where energy use in industry is concerned, energy conservation training finds itself confronted with the many different facets of Production Engineering, with these lodged in many unlikely faculties; for example, in one ancient seat of learning Production Engineering finds itself in the Department of Geography! The fuel supply industries often support research and teaching posts in these faculties, chiefly to promote the interests of their own fuel or power sources; some have funded posts directly concerned with energy efficiency, sometimes closely linked to the Government's BRECSU and ETSU. There is certainly plenty of diversity, which is unlikely to diminish - if, in fact, that is academically desirable - until there is some sort of unified end-product; such as an energy management qualification as a certification requirement for "Energy Manager". If such a requirement were made - as in Japanese industrial law - there would certainly be a scramble among several professions and many university faculties to be the lead body.

12.4.7 The various management professions have been remarkably slow in the U.K. to recognise the opportunities in energy management qualifications. The Institute of Management had a brief flirtation with the EEO over the now-defunct "Energy Managers Workshops" but this developed no further than a one-week course, with neither examination nor certification. Considering that an Energy Manager in any significantly sized industrial, commercial or administrative body requires many management skills, beyond the technicalitics of energy using plant and structures, it is surprising that the Institute of Management has not taken up the challenge, nor even the Institute of Energy. A recent Government (EEO) development suggests that there is some chance of improvement in the matter of higher training for energy management; along with a change in attitude by both Government and Whitehall. This development is the EEO/Management Charter Initiative. The difficulty is that it links energy management training with an initiative that has yet to prove its own credentials and academic integrity and may well sit uneasily in the already complex area of energy training in higher education.

12.4.8 The Management Charter Initiative is an employer-led organisation, backed by the Government and seeking to establish National Standards of best practice for supervisors and managers in the U.K. leading to NVQ/SVQ awards. The cynic might say it is a typically

English ruse to achieve a semblance of qualification without the cost of training and the rigour of examination. However that may be, the EEO has taken up this NVQ/SVQ route to try to get some progress on developing energy management qualifications. The initiative is jointly funded by EEO, British Gas and the Electricity Association, with support from BRECSU, ETSU, academia and the professions. It has yet to be seen whether this route will be any more successful than reliance on further and higher education establishments independently to produce effective training in energy management.

12.4.9 Training the Energy Manager is not, of course, the only objective for energy conservation in higher education. Energy industries, government and quasi-government departments, energy policy consultants and research organisations will be seeking more than "energy literacy" among the graduates they recruit, especially if that energy literacy is confined to the technology of energy use and control. They will be looking for awareness of the economic, political and social issues; of financial aspects at both the macro and micro-level; of the place of energy conservation in merchandising and in the market as a whole; of the psychology and sociology of energy use, ranging from special interests such as Fuel Poverty to the effects of energy needs and environmental effects on whole populations. To date, there is little on offer from academia to produce the rounded energy or energy conservation graduate. Some excellent post-graduate and Masters courses are attempting to do something in that direction but are often technology led. At the other end of the scale, many energy efficiency industries cannot recruit technicians capable of commissioning a Building Energy Management System. There is certainly a lot to be done in both education and training beyond the school curriculum.

13 WHERE NEXT?

Energy conservation has come a long way since Second World War exhortations to save fuel and since the heady days of "Save It", that arose from the various oil crises. In that time it has enjoyed some distinctly different motivations. During Wartime the nation had to face up to living frugally and to denying itself the luxury of wasteful use of an essential commodity that, literally, cost lives to deliver. In the Seventies, the "Save It" campaign was initiated because we feared that the energy supplies were going to run out as the Sheiks closed the oil taps. The infamous "Policy Gap" of official projections raised the spectre of a nation starving, freezing and getting dirty in the dark. From the late Seventies and into the Eighties, we were less worried about having sufficient energy than about being able to afford it, as OPEC actions forced-up World oil prices. "Cost Effective Energy Conservation" became the order of the day and even "conservation" itself became a forbidden word in official circles, to be replaced by the more politically correct "management" or efficiency, (It is curious that the Left is beaten so cruelly with the stick of Political Correctness when the activity was first manifested, on a serious scale, by the Lunatic Right of the Eighties when discussing energy conservation!). The underlying motivations for the changes have been discussed in this book and it should have come as no surprise to those studying the decision processes of energy conservation that the entire movement soon lost momentum over this period. By the mid-Eighties the U.K. had slipped, according to the Secretary of State himself, to the bottom of the world league in energy conservation activity.

As supply pressures had ceased to count, because the UK was now - in the short-term, at least - "energy rich", a far stronger and sinister imperative added to the official willingness to remove energy conservation from the political agenda. This was the needs of the energy supply privatisation programme: At all costs buoyant energy markets had to be preserved to support "The Sale of the Family Silver". Then came the next imperative, "The Environment".

The whole World woke up to the fact that profligate use of energy was a major component in environmental damage. Energy conservation in the U.K. benefited from a whole new impetus but little was achieved. Too much harm had been done, since 1979, to the whole structure of the energy conservation movement for any short-term revival to be possible. Even now, as official flagships, such as the Energy Saving Trust, reveal their fundamental unseaworthiness and as the unofficial bodies remain marginalised, energy conservation in the U.K. is languishing again. However, it is possible to answer the question, "Where Next" with some glimmer of optimism. It is possible to construct a scenario where a new motivation gives renewed impetus to energy conservation and produces something nearer to direction than mere drift. This impetus will come, not from global agents, but from a new pressure that is arising from individuals' health fears.

After worrying that energy would run out, that we could not afford it and then that it was damaging our environment, we may now wake up to the fact that energy use is killing us. That could give quite a new purpose to the business. That rapid and unexpected progress can be made with such motivation was demonstrated by the Lead-in-Petrol saga. Once the health damage was recognised as being a real, and more importantly, universal hazard, both public and government were prepared to act quickly. Threats to middle-class, particularly ruling-class, health and lives are potent motivators for change in the body politic.

In matters of environmental damage, the causal link between energy use and "acid rain" damage has become received wisdom. However, deeper scientific study - and even a little common-sense deduction - reveals a more insidious and personal problem for us all. The energy-produced pollutants do not merely kill trees, invertebrates and vertebrates directly. When subjected to such pollutants, life appears to be extinguished, not so much by the pollutants as by a greater susceptibility to established diseases and parasites. Natural resistance - or, to put it more fashionably, "the immune system" - is damaged by energy pollutants; a sort of energy H.I.V. effect. That is where energy conservation is going next. Peoples and governments will realise that energy use is damaging to personal health. Apart from the "Energy HIV effect", science is producing plenty of evidence about direct health damaging effects at regional and local levels. The fashionable, idle young couple, using the status signifying large car to collect little Samantha half-a-mile from school, will wake up to the fact that such behaviour is killing the little girl. Stand outside any middle-class school at "home time" and you can see the problem now. The pollutants are there to see - collecting at child height. Then there are "estate level" pollution and "indoors air-quality" issues to consider - all arising from the use of energy. Oil, gas and electricity suppliers will realise that the remedies that are needed to meet this life threat mean selling less of their products: not, "Use Wisely", which they hid behind in the past but, "Use Much Less", which has massive and challenging economic problems for all. Even if it means cutting delivered-energy markets by the technically feasible 20%, the health issue is going to insist on such a solution.

That is where energy conservation goes next. For the U.K. it will be even more painful than for most of its E.U. partners. 15 years of neglect and political vandalism have left the U.K.'s energy conservation capability in poor shape and the energy industries fragmented and commercially uncontrolled. On the other hand, that should be an exciting challenge for a new and , we hope, angry generation.